EARLY EDITIONS OF EUCLID'S ELEMENTS

BY

CHARLES THOMAS-STANFORD

SAN FRANCISCO
ALAN WOFSY FINE ARTS
1977

The first edition of this work was published in 1926 by The Bibliographical Society in London as Volume No. XX in the *Illustrated Monographs* series.

This new edition has been published by arrangement with The Bibliographical Society.

Alan Wofsy Fine Arts

150 Green Street

San Francisco, CA 94111

This edition is limited to 450 copies, of which 134 contain an original leaf from the first edition of Euclid's *Elements* (E. Ratdolt, Venice, 1482).

ISBN 0-915346-29-X

ISBN 0-915346-30-3 (with leaf)

LC No. 77-7448

Printed in the United States of America

PREFACE

I WAS led to collect and to study the early editions of Euclid's Elements by the beauty of some of the books, and by the interest attaching to the ingenuity of printers in associating a continuous series of diagrams with the text. I found that the bibliographical literature relating to these editions was meagre, and that the great libraries were rather deficient in them—a deficiency which has been considerably diminished at the British Museum since I called attention to the subject. From the outset I was encouraged in this pursuit by the interest of the late Mr. Francis Jenkinson, who supplied me with details of Euclids in the University Library, Cambridge. At a later date Mr. A. W. Pollard was good enough to do me the same service as regards the British Museum. My best thanks are due to Mr. F. G. Rendall, who has carried out my scheme for a bibliography with great skill and thoroughness.

C. T.-S.

CONTENTS

LIST OF NEW PLATES

LIST OF NEW PLATES

All of the new plates have been reproduced from the original editions at the *Bancroft Library, University of California, Berkeley,* with the exception of the Cavellat edition of 1557 which is at the *San Francisco Public Library.*

EARLY EDITIONS OF EUCLID'S ELEMENTS

1482—1600

LITTLE is known of Euclid personally. According to Proclus, who wrote an elaborate commentary on the first book of the Elements, he flourished at Alexandria about 300 B.C. In the early editions he is commonly and erroneously described as Megarensis—a confusion with a Socratic philosopher of the same name born at Megara in the fifth century B.C. The history of Greek Mathematics, and of Euclid's place in it, is told clearly and concisely by Sir Thomas Heath in a book of essays by different writers entitled *The Legacy of Greece* (1922). Of the Elements he says: 'No work presumably, except the Bible, has had such a reign; and future generations will come back to it again and again as they tire of the variegated substitutes for it, and the confusion arising from their bewildering multiplicity.' My concern is not with such high matters as the importance of Euclid in the evolution of geometry, or even with the critical value of the numerous editions of the Elements in various languages, which began with Ratdolt's Latin Edition in 1482; it is of course rather with those details of date, type, illustration, editorship, and so on, which we describe as bibliographical.

These editions of the fifteenth and sixteenth centuries are for the most part uncommon books; some, especially the more insignificant, are of great rarity, perhaps because they were largely used by schoolboys. At the great national libraries, the British Museum and the Bibliothèque Nationale, students seem to be less completely supplied with them than with other series of works of like importance; the University libraries are singularly deficient considering the nature of the books. The great private libraries which have been dispersed during the past century, from that of Dr. Kloss downwards, seldom contained any but Ratdolt's edition and the Greek *editio princeps* printed at Basle in 1533.

On the other hand, the sale catalogue of the Pinelli library, dispersed in 1789, mentions seven editions earlier than 1600, and the catalogue of the historian De Thou's library published in 1679 contains eleven such editions. His copy of the Roman issue of 1574 in two volumes, with his bachelor arms on the cover, is in my possession.

The only modern bibliography of Euclid with which I am acquainted is Professor Pietro Riccardi's Essay towards a bibliography published at Bologna in 1887 and following years. It is not very helpful. It is only a hand-list with occasionally a few bibliographical details; many supposed editions of which no trace can now be found are included on the doubtful authority of early bibliographers, and many important editions are omitted. As it reaches down to recent times it of course covers a wide field, and it may perhaps be more valuable in the later periods than I have found it in the earlier.

For convenience I propose to divide the books which come within my limited period into four classes:

I. The 'full-dress' editions of the complete Elements, or of not less than the first six books; with the demonstrations and with diagrams; in Greek, Greek and Latin, or Latin.

II. Editions in Greek or Latin of the enunciations only, without the demonstrations; with or without diagrams.

III. Translations into current European languages—Italian, French, German, English, and Spanish; also a translation into Arabic made at an earlier date.

IV. Fragmentary editions in various languages.

Of these the first and third are by far the most important. The existence of Class II is mainly due to the erroneous belief current in the fifteenth and sixteenth centuries that Euclid only left the definitions, postulates, and enunciations; and that the demonstrations were the work of Theon the Younger of Alexandria who lived in the fourth century A.D., or of some later commentator. This belief survived until the time of Sir Henry Savile, when the view now commonly held was established—

that Euclid was the author of the demonstrations also. The words of Proclus, who lived shortly after Theon, in his commentary on the first book of the Elements, seem to me to give powerful support to this view: 'Ea quae a prioribus molliore brachio ostensa fuerant ad eas redegit demonstrationes quae nec coargui nec conuinci possunt.' The matter is fully gone into by the late Professor De Morgan in his article on Euclid in Smith's *Dictionary of Classical Biography.*

The fragmentary pieces—sometimes translations of single books for special purposes, and so on—which form Class IV, perhaps do not strictly fall within the scope of this paper at all; but in any full bibliography of the Elements they would have a place. The most important unit in this class is the *Geometriae libri duo* of Boethius, the first European translation of any part of Euclid. It is a very small part.

Works merely based on Euclid are not within my purview. I regret that this rules out the geometrical treatise of Archbishop Thomas Bradwardine, printed at Paris in 1495 and frequently reprinted in France and Spain. It is avowedly derived from Euclid, but it is not Euclid.

The first printed edition of Euclid appeared, as I have mentioned, in 1482 from the press of Erhard Ratdolt at Venice. It was an epoch-making book from many points of view. It was the first attempt—and a highly successful one—to produce a long mathematical book illustrated by diagrams.[1] In his dedicatory epistle to the Doge Ratdolt says that he had often wondered why the great output of the Venetian printing presses included so few and so insignificant mathematical works. He had discovered that this was due to the difficulty of printing the diagrams without which such books were almost unintelligible, and he had set himself to overcome the difficulty. He certainly did overcome it; and he did more. He not only produced a very beautiful book, but he set a standard for the production of mathematical books which lasted for a generation or longer. I may instance not only later Venetian editions of Euclid, but some of the work of Hamman alias Herzog in the fifteenth century (such as his *Regiomontanus*), and of

[1] It seems probable that an important book with mathematical diagrams preceded, or was contemporary with, Ratdolt's edition of Euclid. Of three books printed by Petrus of Corneno at Milan, the first was dated 31 July 1480; the second 30 May 1481; the third, the *Perspectiua Communis* of Archbishop Peckham, is undated. The *Perspectiua*, a folio of thirty leaves, contains seventy-seven complicated and well-executed diagrams.

Köbel at Oppenheim in the first quarter of the sixteenth. It would almost seem that at Venice especially the printers sought by a refinement of ornamentation to relieve the austerity of the subject-matter.

The Latin version printed by Ratdolt is said to have been made by Adelard of Bath in the thirteenth century, whether from the original Greek or from an Arabic version of it appears to be uncertain. It is stated that the Euclidean text is accompanied by the commentary of Campanus of Novara. Adelard of Bath is a striking personality of the later Middle Ages, an embodiment of the new spirit which in the twelfth century was animating the Universities, both of England and the Continent. 'At the Court of Henry I. might occasionally be seen the figure of Adelard of Bath, the "English philosopher" who had travelled to Italy, Sicily and Asia Minor, in quest of scientific information. As the author of a dialogue, *De rerum causis*, and as the translator of Euclid into Latin, he holds a definite though humble place among those who restored the intellectual contact of the west with the east, of medieval learning with ancient science.' [1]

The type used in Ratdolt's Euclid is gothic, 45 lines to the page. The book contains 138 leaves. The first page of the text has a very beautiful border, already used in the second volume of Ratdolt's *Appian*, and there are many hundreds of woodcut initials of Ratdolt's well-known series. Mr. Redgrave states in his monograph of Ratdolt that there are upwards of 420 woodcut illustrations, excluding about 200 which may be formed from lead lines. But there seems to be some doubt whether the diagrams are woodcut. De Morgan says, 'though at first sight they seem to be woodcuts, yet a closer inspection makes it probable that they are produced from metal lines'.

There are copies in which the dedication is printed in gold, two of them being on vellum, one in the British Museum, the other in the Bibliothèque Nationale. Mlle Pellechet notes in these some minor differences from the ordinary issue in the make-up of the text.

So much for Ratdolt's *editio princeps*. In contemplating the finished perfection of his work, the work of an artist who could make of the diagrams set in the wide margins of his book an added beauty to it, it is difficult to remember that it was also the work of a pioneer. As an

[1] H. W. C. Davis, *England under the Normans and Angevins*, 1921, p. 184.

explorer may hew a way through a virgin forest and thereby determine the course of a road which will become permanent, so the great printer found a style and laid down a method which were followed by his successors for centuries.

In 1491—nine years after the appearance of Ratdolt's book—a reprint of it came from the press of Leonardus Achates and his partner Gulielmus of Pavia at Vicenza. It is curious that though it seems to be an exact page for page reprint the type used is roman instead of gothic. Mr. Stephen Gaselee has called my attention to the date in the colophon. It is given as 'Anno Salutis 1491 Vigesimo Calen. Iuñ.' This ignoring of the Ides as the proper basis of dating in such a case is due I think to slavish copying of Ratdolt. He had given the date of his edition as 'Anno salutis 1482 Octauis. Caleñ. Iuñ.', and the Vicenza printers wishing to date theirs twelve days earlier in the month substituted 'vigesimo' for 'Octauis'.

The book has a border to the first page which is of an entirely different character from Ratdolt's. Instead of floral interlacements there are semi-grotesque animals, birds, and cupids. Individually some of the initial letters are good, but they are very miscellaneous and belong to different sets. The diagrams often lack the precision of the earlier edition, and have much more the appearance of woodcuts than Ratdolt's. But tried by a less exacting standard of comparison it is a fine book, and it may be accounted much more uncommon than the *editio princeps.*

Although it does not fall within this class, I may mention in passing that an edition of the first four books was printed at Frankfort in 1506. The only copy I have seen is in the British Museum; it appears to be a reprint, as far as it goes, of Ratdolt's edition.

Meantime a very important edition had appeared in 1505 printed at Venice by Joannes Tacuinus. It was a translation into Latin from a Greek text by Bartolommeo Zamberti, who claims that he has restored and excluded from the exposition of Theon many things that were 'subuersa et prepostere voluta' in the version of Campanus. For example, the Pythagorean proposition becomes the 47th of the first book as we know it. Zamberti contributes a long preface and a life of Euclid. The thirteen books of the Elements are followed by the Phaenomena, Specularia, &c. The volume itself is a first-rate example of the Venetian book of the time.

There is an elaborate title-page with the printer's well-known cut of John the Baptist at the foot. The first page of the text has a fine border, and the larger initial letters are a charming set depicting children playing. The enunciations are printed in gothic, the demonstrations in roman type.

In 1510 some of the same sheets were reissued with a freshly printed last page containing the colophon, a misreading of which has often led to the issue being described as of 1517. Both issues seem to be among the rarest of early Euclids. I have both, as has the British Museum; the Bibliothèque Nationale has only that of 1510, the Bodleian and the University Library, Cambridge, neither.

Meanwhile in 1509 there had appeared also at Venice a very notable edition of the fifteen books of the Elements from the press of Paganinus de Paganinis (called A. Paganius Paganinus on the title-page). From the typographical point of view it is a very remarkable and attractive book. The title, in red and black, is admirably spaced. The text, which is not overloaded with commentary, fills only half the width of the page, the ample margin being occupied by the diagrams which are on an unusually large scale. The enunciations are printed in gothic, the demonstrations in a diminutive but clear roman (or semi-roman) type, of which 20 lines measure 72 millimetres. The editor, Lucas Paciolus, better known as Lucas de Borgo, is very severe on previous editions 'quae cum antea librariorum detestanda culpa mēdis fedissimis adeo deformia essent vt vix Euclidem ipsum agnosceremus'.

A curious feature of the volume is that a lecture on the fifth book given by the editor in the church of St. Bartholomew at Venice on the 11th August 1508 is prefixed to that book, and an imposing list of the eminent persons present is added after the manner of the modern newspaper report. It occupies a whole page. I notice the name of Aldus Manutius Romanus. It may be remarked in passing as rather strange that no edition of Euclid was printed by the Aldine press during the century of its existence.

In the Venetian Euclids of 1482, 1505, and 1509 the art of book-production reached the meridian. Subsequent editions, produced elsewhere, of varying typographical quality, seem to be more distinctly commercial in character and to be destined for customers to whom utility and perhaps cheapness were main considerations. And it would seem that having produced these masterpieces Venice did not attempt subsequently to compete with Paris, Basle, Strasburg, and other towns. After 1510 no Euclid was printed at Venice, except an Italian translation.

The first edition printed in France came from the press of Henri Estienne in 1516, a folio in roman type. It contains the commentaries (so called) of Campanus Theon and Hypsicles. The diagrams are well executed, but the tradition of the book beautiful is not maintained. Euclid is still styled Megarensis. The editor was Jacques Faber, or Lefebre, surnamed Stapulensis from his birthplace (Etaples), a scholar of the period and a pioneer of the Protestant movement in France. There are slight variations in different copies of this edition.

I now come to the Basle editions, an important series, all folios, printed by Johann Herwagen between 1533 and 1558. Herwagen had migrated from Strasburg about 1528, when he acquired the citizenship of Basle, and married Gertrude, widow of the learned Basle printer John Froben and daughter of the scholar and patron of letters Wolfgang Lachner. He was the first printer to inset Euclid's diagrams in the text. Earlier printers, and some later, placed them in the fore margin. His editions begin with the *editio princeps* of the Greek text, published in the month of September 1533. It is printed in a small much-contracted greek type, 47 or 48 lines to the page, of the sort that almost makes one weep when one remembers what greek type may be—as in the work of Lorenzo di Alopa in Italy, or Arnaldo de Brocar in Spain. The Elements occupy 268 pages, followed by 115 pages containing the four books of the commentary of Proclus on the first book of the Elements, which according to Graesse has never been reprinted. A Latin version was printed at Padua in 1560. The editor of the Greek text of 1533 was Simon Gryner or Grynaeus, Professor of Greek at the University of Basle, who supplies an introduction (in Latin) in the form of a dedication to Cuthbert Tunstall, Bishop of Durham. Gryner had visited England two years before, and had been warmly received by Sir Thomas More on the introduction of Erasmus. Doubtless he had met Bishop Tunstall on that occasion. The text was edited from two manuscripts sent to Gryner, one from Venice by Lazarus Bayfius, the other from Paris by John Ruellius; the commentary of Proclus from a manuscript supplied by John Claymond, President of Magdalen and later of Corpus

Christi College, Oxford. Gryner describes his friends as 'viri optimi et humanissimi: literis iuvandis et exornandis facti: quod ipsorum monumenta docent'.

The volume is a typical example of Basle book-production of the time, business-like and lacking distinction—a disappointing book. One cannot but feel that Herwagen missed a great opportunity. No doubt he did not foresee how great an object of interest the *editio princeps* of this important Greek text would be to men born many centuries after its production.

Four years later, in August 1537, he published a Latin version of the Elements, followed by other works attributed to Euclid. It is in roman type and contains three pages of introduction by Philip Melanchthon addressed 'studiosis adolescentibus'. From many copies this introduction has been removed by the clerical censor who has added his stamp. As there does not appear to be anything objectionable in the introduction itself, this action of the censor must have arisen from hostility to the writer of it. From my copy the three leaves have not been removed.

A reprint of this edition appeared in 1546; it is quite differently set up. From my copy the introduction has been removed. Another reprint was published by Herwagen and Bernard Brand, his son-in-law, in 1558. One copy I possess of this edition has the title-page of 1558 with the colophon page of 1546, and appears to be partly made up of sheets of that edition.

Previously, in September 1550, there had appeared from Herwagen's press an edition of the first six books of the Elements. The enunciations are printed in Greek and Latin. The editor was Johan Scheybl, professor of mathematics at Tubingen. The text occupies 315 pages, of which the first 76 contain a 'brief description of the rules of Algebra' by the editor. It is a handsome book, with the diagrams on a larger scale and printed on better paper than the other Basle editions. The only copies I know of, other than my own (formerly in Lord Chesterfield's library), are those in the Cambridge University Library, British Museum, and Bodleian Library.

In following the Basle editions I have strayed from the chronological order, and it is necessary to retrace my steps.

In the second quarter of the sixteenth century there flourished at Paris a professor of mathematics called Fine Oronce (Finaeus Orontius on his title-pages). He wrote many books on arithmetic, geometry, and astronomy; and is curtly described by Mr. David Eugene Smith of Columbia University, author of *Rara Arithmetica* (a book which, as far as it touches my subject, I have not found very accurate), as 'not a genuine scholar'. That may be so; but that was not the opinion of the time, when he had a great reputation and his books a great vogue. In 1536 Simon de Colines printed his edition of the first six books of the Elements. The enunciations are printed in Greek and Latin; the demonstrations are set out by Oronce. They appear to be based on the Venetian edition of Zamberti. The volume is an attractive one to the eye, as is much of Simon de Colines' work. There is an elaborate border to the title representing the mathematical sciences. It was reprinted in 1544 with a more refined border of interlacements on a criblé ground in the best style of the period. It is a fine book: the Venetian tradition is not yet dead. Another reprint appeared in 1551.

At Lyons a folio edition of the first six books, edited with a revision of the demonstrations by Jacobus Peletarius (Jacques Peletier), was printed by Joannes Tornaesius (Jean de Tournes) and Gulielmus Gazeius; at the end are appended several letters of the editor to his friends. It has an emblematic border to the title-page, and is quite a handsome book.

Another folio edition of the first six books appeared at Strasburg in 1566 from the Rihel press, stated to be 'pro schola Argentinensi'. The enunciations are printed in Greek and Latin, the demonstrations and commentary in Latin. The editor of the first and fifth books was Christianus Herlinus, of the others Conrad Dasypodius alias Rauhfuss, the maker of the celebrated Strasburg clock which was constructed in 1574 and lasted until 1789. The Rihel press had previously printed, in 1559, a small octavo volume of 27 printed leaves containing the first book only in Greek, stated in the preface to be specially for the use of those who wished to understand the Dialectical books of Aristotle. In the succeeding twenty years Dasypodius brought out a large number of school books which fall into Class IV.

The year 1566 likewise saw the appearance at Paris, from the Royer press, of a more important edition containing the fifteen books with the addition of a sixteenth, *De Solidis*. Its editor was François de Foix,

Comte de Candale, Bishop of Aire, who makes the not unusual claim that he has eliminated numerous errors into which his predecessors had fallen. This edition was reprinted more than once.

So far only the chief book-producing centres had been concerned in the printing of Euclid. A provincial town of Italy now took the torch.

A new translation into Latin of the fifteen books of the Elements appeared at Pesaro, a seaport on the Adriatic near Urbino, in 1572. The printer was Camillo Francischini. The translation, which was made use of by subsequent editors for centuries, was the work of Federigo Comandino, certainly an outstanding figure in the history of Euclid's Elements. He was born at Urbino in 1509 of a noble family. Successive dukes of Urbino for a century and a half combined the pursuit of arms with the patronage of art and science. The theory of fortification and its practical application to the fastnesses of Umbria led to the study of pure mathematics. In the atmosphere of this court young Comandino was brought up.

'After a liberal education he passed several years at the Court of Clement VII, nominally as a privy chamberlain but really to amuse with learned disquisitions the Pontiff's leisure hours, on whose death he repaired to Padua, where he devoted ten years to the study of philosophy and medicine. Having graduated, he settled for clinical practice at Ferrara, but seems soon to have abandoned the healing art for mathematical research. He accompanied his sovereign, Guidobaldo II, to the camp at Verona, when in the Venetian service, and having gained his confidence by successfully treating him in a severe illness, he was selected to instruct him in astronomy and cosmography. . . . In studious retirement he devoted to the exact sciences the matured powers of a comprehensive and most retentive mind. He explored all that classical authors were known to have left on these subjects, and rendered again accessible much that lay forgotten among the rubbish of by-gone learning. . . . After spending the prime of life in these pursuits at Rome, he returned to his native duchy, where his instructions in mathematics were sought by Prince Francesco Maria, with whom he read and expounded Euclid's Elements; and afterwards at the request of his pupil, published a Latin translation of them. It was about 1569 that he was visited there by a young Englishman named John Dee, whose love of the exact sciences induced him to seek so distinguished a professor, and who supplied him with some Arabic manuscripts hitherto unknown. Six years thereafter he was surprised by death, with many unfinished works on his hands. . . . The life of a hard student is rarely one of varied incident; and even the voluble pen of his pupil Baldi has failed to illustrate that of Comandino with interest, beyond his scholiast labours. Yet severity formed no part of his social character, and he was ready at all times to relax his toils by Epicurean indulgences, which are said eventually to have curtailed his life. To the last, however, his engrossing pleasure was in books.'[1]

Shortly before his death an Italian version of his translation was printed at Urbino, to which I shall refer in due course. The Pesaro edition is a folio, with well-executed diagrams, and contains 14 preliminary leaves and 255 leaves of text.[2]

In 1574 a new departure in the matter of format was taken. A Latin edition in two octavo volumes with rather diminutive diagrams was printed at Rome by Vincenzo Accolti. The translator was Christophorus Clavius (Schlüssel) of Bamberg, of the Society of Jesus, a mathematician who gave the Gregorian Calendar or New Style its present form and made all the calculations necessary for its verification. It contains the fifteen books with very full scholia, and the addition of a sixteenth, *De solidorum regularium comparatione.* It was reprinted at Rome in 1589 in the same form, and in folio at Cologne in 1591.

It is unnecessary to deal at any length with the few and unimportant books which compose my Class II—little octavo volumes containing only the enunciations. We may suppose that they were intended for use as adjuncts to professors' lectures—as indeed is suggested in the preface to one of them, and according to the erroneous view of the time, as I have said, they were considered to represent the whole of Euclid's text. The first was printed, in Greek, without diagrams, by Antonio Blado Asolano at Rome in 1545. Greek printing at Rome had fallen to a low ebb. It is stated by Mr. Horatio F. Brown[3] that when Paul III and Cardinal Marcello Cervini desired to establish a Greek press at Rome, they found it necessary to send the Apostolic printer, Antonio Blado, to Venice to find the fount of type. As Alessandro Farnese was elevated to the papacy as Paul III in 1545, this little book was perhaps a trial essay with the new fount. It is sometimes accompanied by an Italian translation by Agnolo Caiani, printed at the same press in the same year. Both books have a portrait of Euclid on the title-page. It may be interesting to note that a copy of the Italian

[1] Dennistoun's *Memoirs of the Dukes of Urbino*, 1909, vol. iii, p. 260.

[2] Mr. Stephen Gaselee informs me that he has seen in a private library a copy printed on blue paper.

[3] *The Venetian Printing Press*, 1891, p. 100.

version, alone, was in the Pinelli sale in 1789. It was described in the catalogue as 'liber perrarus', and was sold for a shilling.

In 1549 Thomas Richard printed at Paris the enunciations of fifteen books without diagrams. The preface states that the object was to save expense, the cost of the earlier editions being greater than the slender means of students could bear. In the same year an edition of the enunciations in Greek and Latin appeared at Leipzig with neat little diagrams; and in 1557 and 1558 Guillaume Cavellat printed at Paris editions somewhat similar, the diagrams curiously decorated with floral ornaments. In the last year of the century the Latin text of Cavellat's editions was reprinted at Cologne, the diagrams, including the decorations, being almost exactly copied.

We now come to Class III, the interesting group of translations into other languages than Latin. Italy again takes the lead. The first Italian translation of the fifteen books of the Elements appeared at Venice in 1543 from the press of Venturino Roffinelli. The translator was Nicolo Tartalea of Brescia, an eminent mathematician of the period. It is a folio volume of no special distinction; the diagrams rather roughly drawn, apparently on wood. On the title-page are the arms of Gabriele Tadino, to whom it is dedicated, and at the end the mark of Pietro di Facolo, bookseller at Venice, by whom it was published. This translation was reprinted at Venice in quarto form by Curtio Trojano in 1565, by Giovanni Bariletto in 1569, and in 1585. In these editions the diagrams seem to be engraved on metal.

Under Class I, I spoke of the Latin translation of Comandino printed at Pesaro in 1572. In 1575 an Italian version of this appeared at Urbino from the press of Domenico Frisolino. It is a folio printed in similar style to that of the Pesaro book—apparently from the same founts of type, and with the same initial letters and diagrams. The privilege of Pope Gregory XIII to Comandino which was prefixed to the Latin edition of 1572 is printed here translated into Italian. It seems rather surprising that such a privilege, conferring a copyright for ten years of a specified translation into Latin, should be claimed to cover a version in another language printed elsewhere three years later.

A French translation of the first six books by Pierre Foscadel de Bezies, Mathematical Reader in Ordinary to the King in the University of Paris, was printed at Paris by the firm of De Marnef and Cavellat in the year 1564. It is a quarto, well turned out in the good tradition of the De Marnef family. The diagrams are the same as in the little edition of Cavellat in Class II, and have the same curious floral ornaments. It is perhaps incorrect to describe it as a full edition of Euclid's first six books. Euclid's enunciations are printed, but the demonstrations or commentaries appear to be by Pierre Foscadel, and are perhaps derived from his lectures. An unusual feature is that while there is a general dedication to King Charles IX, each of the books subsequent to the first is preceded by a letter to some eminent person, of whom the most noteworthy is Gaspard Coligny, the great Huguenot leader and the first victim of the massacre of St. Bartholomew. In some copies the date on the title-page is incomplete, the final numeral being omitted. The introduction is dated April 1564. A similar version of the seventh, eighth, and ninth books was published in 1565.

The first German translation of Books I to VI, a folio of 106 leaves, appeared in 1562 from the press of Jacob Kündig at Basle. It had been preceded by a translation of Books VII, VIII, and IX by Scheubel printed at Augsburg in 1555. Kündig's volume is a rather handsome one, with an elaborate title-page, printed in red and black. It has good woodcut ornaments and well-executed diagrams. It contains a long preface by the translator, Wilhelm Holtzman.

In England the first Euclid to be printed was an English version of the fifteen books of the Elements from the press of John Day in 1570. Robert Recorde's *Pathway to Knowledge*, published in 1551, and again in 1574 and 1602, contains the enunciations of the first four books, not in Euclid's order. Recorde attributes the demonstrations to Theon. To the 1570 book must therefore be allowed the honour of being the first English Euclid. The translator was Henry Billingsley, citizen of London. The old mistake of describing Euclid as Megarensis is repeated. There is some difference of opinion as to the original from which the translation was made. De Morgan says it was 'certainly made from the Greek and not from any of the Arabico-latin versions, and is for the time a very good one'. The *Dictionary of National Biography*, on the other hand, says 'his original was the Latin version of Campanus'.

Billingsley's career was a remarkable one. The son of Robert Billingsley of Canterbury, he was admitted Lady Margaret scholar of St. John's, Cambridge, in 1551. He is said to have studied for several years at Oxford also; but he took no degree at either university. Anthony Wood states that at Oxford he acquired a taste for mathematics under the tuition of 'an eminent mathematician called Whytehead', who was at one time 'a fryar of the order of St. Augustine'. Billingsley was afterwards apprenticed to a haberdasher in the City of London, and became a wealthy merchant. His success might be used as an illustration of the value of a university education as a preliminary to a commercial career, of which we hear something nowadays. He was Lord Mayor in 1597, and Member of Parliament for the City in 1603. He died in 1606.

The following extract from the address of the Translator to the Reader exhibits the scope of the book:

'Many other artes also there are which beautifie the minde of man: but of all other none do more garnishe & beautifie it, then those artes which are called Mathematicall. Unto the knowledge of which no man can attain, without the perfecte knowledge and instruction of the principles, groundes, and Elementes of Geometrie. But perfectly to be instructed in them, requireth diligent studie and reading of olde auncient authors. Amongest which, none for a beginner is to be preferred before the most auncient Philosopher Euclide of Megara. For of all others he hath in a true methode and iuste order, gathered together whatsoeuer any before him had of these Elementes written: inuenting also and adding many thinges of his owne: wherby he hath in due forme accomplished the arte: first geuing definitions, principles, & groundes, wherof he deduceth his Propositions or conclusions, in such wonderfull wise, that that which goeth before, is of necessitie required to the proufe of that which followeth. So that without the diligent studie of Euclides Elementes, it is impossible to attaine unto the perfecte knowledge of Geometrie, and consequently of any other of the other Mathematicall sciences. Wherefore considering the want & lacke of such good authors hitherto in our Englishe toungue, lamenting also the negligence, and lacke of zeale to their countrey in those of our nation, to whom God hath geuen both knowledge, & also abilitie to translate into our tounge, and to publishe abroad such good authors, and bookes (the chiefe instrumentes of all learninges): seing moreouer that many good wittes both of gentlemen and others of all degrees, much desirous and studious of these artes, and seeking for them as much as they can, sparing no paines, and yet frustrate of their intent, by no meanes attaining to that which they seeke: I haue for their sakes, with some charge & great trauaile, faithfully translated into our vulgar tōge, & set abroad in Print, this booke of Euclide. Whereunto I have added easie and plaine declarations and examples by figures, of the definitions. In which booke also ye shall in due place finde manifolde additions, Scholies, Annotations and Inuentions: which I haue gathered out of many of the most famous & chiefe Mathematiciēs, both of old time, and in our age: as by diligent reading it in course, ye shall well perceaue.'

Preceding the text of Euclid is a lengthy Preface by the celebrated Dr. John Dee (1527–1608), whom we have already heard of at Urbino. Dee was a mathematician, an astrologer, a collector of books and manuscripts, and a traveller; a Fellow of Trinity College, Cambridge, where the clever stage effects which he introduced into a performance of the *Peace* of Aristophanes procured him a lifelong reputation as a magician, a reputation which he bitterly resented. Towards the end of this Preface, which ranges over the whole field of mathematics, he breaks off into a passionately indignant digression, a brilliant piece of florid invective against his maligners.

'For marueilous Actes and Feates', he asks, 'Naturally, Mathematically, and Mechanically, wrought and contriued: ought any honest Student, and Modest Christian Philosopher, be counted, and called a Coniurer? Shall the folly of Idiotes, and the Mallice of the Scornfull, so much preuaile, that He, who seeketh no worldly gaine or glory at their handes: But onely, of God, the threasor of heauenly wisedome, and knowledge of pure veritie: Shall he (I say) in the meane space, be robbed and spoiled of his honest name and fame? He that seketh (by S. Paules aduertisement) in the Creatures Properties, and wonderfull vertues, to find iuste cause, to glorifie the Æternall, and Almightie Creator by: Shall that man be (in hugger mugger) condemned, as a Companion of the Helhoundes, and a Caller, and Coniurer of wicked and damned Spirites?'

He declaims against the 'common peeuish Prattler' and the 'malicious Scorner' who have libelled him, and his unkind unthankful countrymen who have oppressed him; and he declares that he is 'innocent in hand and hart'.

At the end of Book X is an Advertisement by John Dee touching the last proposition of that book.

Perhaps the seeming incongruity of a haberdasher and alderman producing so notable a book has led some to doubt the fact. It has been suggested that Dee was really the translator. But his turgid style would surely have asserted itself in the editorial notes, in which I find no trace of it. Others point to the fact that Whytehead, his old tutor, passed his declining years in Billingsley's house, and without dimming Billingsley's fame we may fairly suppose that the old mathematician assisted the busy man of affairs in his literary work.

The book itself is a remarkable production, a stout folio in the well-known manner of John Day whose portrait trademark fills the last page. Apart from the introductory matter there are 930 pages of text with diagrams well executed. The title-page, from a block which had been used by Day earlier, is elaborately emblematic, containing figures of various philosophers (not including Euclid), and of females representing Geometry, Arithmetic, Astronomy, and Music. At the foot is Mercury. I am aware that early Greek sculptors sometimes endowed the youthful Hermes with a nascent beard, 'the doubtful down' that is the promise of the man to be. But here we have the sedate figure of a corpulent and full-bearded citizen of middle age which is very unmercurial. If the block were not of earlier date it would be tempting to suppose that in this figure of Mercury the Messenger we had a portrait of Billingsley the Translator.

It is a descent from this really notable English book to the insignificant Spanish version printed in 1576. That no edition of Euclid either in Latin or Spanish had appeared in Spain before is rather strange! Mathematical books with diagrams had been printed in Spain at a much earlier date. Archbishop Bradwardine's *Compendium of Geometry* was printed at Valentia by J. Jofre in 1503. It was included in the *Cursus quattuor Mathematicarum artium liberalium* of Pedro Ciruelo printed at Alcala by de Brocar in 1516, and again in the Complutensian University in 1526–8, a work which contains extremely elaborate diagrams, especially in the section relating to Perspective.

As regards Euclid, the edition of 1516 catalogued by the credulous Riccardi on the authority of Graesse is certainly fictitious. It is doubtful whether the editor was born at the time. In Graesse's book the date 1516 is probably a misprint for 1576. The edition of this year, a translation of the first six books by Roderigo Zamorano, Astrologer, Mathematician, and Cosmographer, was printed at Seville in the house of Alonso de la Barrera.

It is a small quarto of 122 leaves and has the appearance of a school book. Such a use would account for its rarity and the rather worn condition of the copies I have seen.

The last book I have to mention in Class III is the most remarkable—possibly the most remarkable of all printed editions of Euclid. I ought perhaps to have put it in a class by itself. It is not a contemporary translation into the vernacular of any European country. It is the Arabic version of Euclid's thirteen books made by the astronomer Nāṣir al-dīn in A.D. 1260, and printed at Rome at the Medicean press in 1594. Printed throughout in Arabic, it is a distinguished product of the Roman press, and a typographical monument of which any press might be proud. In the copy I possess the only indication of European origin is the numbering of the leaves in small Roman numerals at the foot of the page. Other copies have a Latin title-page. Some contain only twelve books and have 400 pages instead of 453.

Little need be said of Class IV. The most important work which falls within it is the translation into Latin of portions of the earlier books of the Elements attributed to Boethius. This is contained in *Opera Varia Boethii*, a noble volume printed at Venice by Joannes and Gregorius de Gregoriis in 1491–92, and was often reprinted, frequently in conjunction with the *Sphaera Mundi* of Johannes de Sacro Bosco. It is entitled *Euclidis Megarensis Geometricae libri duo* and contains most of the definitions of the first three books of the Elements, the postulates, and most of the axioms. The enunciations are given, with diagrams, but no proofs. Boethius was born about A.D. 473, and died about 524. The correctness of the attribution to him of this work has been questioned, but it is probably in any case the first translation of any part of Euclid into a European language.

A long series of school books edited by Dasypodius and printed at Strasburg round about 1570 is an important feature of this class.

Portions of Euclid lie hidden in treatises and lectures of the sixteenth century, and I cannot hope that all such have been discovered and included in the Bibliography of this section. Their discovery may be accidental. For example, it was quite by chance that in turning over the leaves of Bartoli's *Perspectiua* (Venice, 1560) I lighted upon a portion of the

Elements translated into Italian. The collection of mathematical papers by Maurolico (Venice, 1575), containing the thirteenth book of the Elements, is another case somewhat of the same kind, although here the table gives an indication of the contents.

An effort has been made to render the Bibliography of this section as comprehensive as possible, and it may be perhaps considered that in some instances it passes beyond the strict limitations of my subject; it is not easy to decide where exactly to draw the line. But I fear that nevertheless the Bibliography of Class IV must be regarded as rather tentative than complete.

BIBLIOGRAPHY

I

EDITIONS OF NOT LESS THAN THE FIRST SIX BOOKS WITH DEMONSTRATIONS, ETC., IN GREEK, GREEK AND LATIN, OR LATIN

1.

1482, 25 May. Venice. Erhard Ratdolt.

1^b. Erhardus ratdolt Augustensis impres|sor Serenissimo Almae vrbis venetae | Principi Ioanni Mocenico .S. 2^a. (red) Preclarissimū opus elementoꝝ Euclidis megarēsis vna cū cō-|mentis Campani ꝑspicacissimi in artē geometriā incipit felicit. 19^a. Propositio . 18. | SI circulum linea recta contingat: . . . 137^b. ℭ Opus elementorū euclidis megarensis in geometriā artē In id quoqꝫ Campa-|ni ꝑspicacissimi Cōmentationes finiūt. Erhardus ratdolt Augustensis impressor | solertissimus. venetijs impressit. Anno Salutis. M.cccc.lxxxij. Octauis. Caleñ. | Iuñ. Lector. Vale.

Folio. a^{10} b–r^8. 138 leaves, the last blank. 3^a: 45 lines (mixed types), 202 (with headline 211) × 120 (with diagrams 187) mm. Types: gothic throughout except headlines (book-numbers) roman. Ornamental capitals, capital space on 2^a. Marginal diagrams.

Some copies have the dedicatory letter printed in gold (e.g. George III's vellum copy now in the British Museum, the vellum copy in the Bibliothèque Nationale, and according to Redgrave, *Erhard Ratdolt*, that in Augsburg City Library).

B.M.

1 a.

A variant.

1^b. ℭ Erhardus ratdolt Augustensis impressor. Serenissimo | alme vrbis venete Principi Ioanni Mocenico. . S. 2^a. (red) Preclarissimus liber elementorum Euclidis perspi-| cacissimi: in artem Geometrie incipit quāfoelicissime. 19^a. Propositio . 18. | SI circulū linea recta cōtingat: . . .

In this variant leaves 1–9 are differently set up. The heading on 1^b is printed with text type, not with heading type as in the issue first described, and a three-side border occurs on 2^a. Errors on the two outer pages of sheet c 1 have been corrected.

Hain *6693. Proctor 4383. Redgrave 26. B.M.C. Pt. v, pp. 285, 286. Riccardi.
B.M. Bodl. C.T-S.
See pl. 1, 2.

2.

1491, May. Vicenza. Leonardus de Basilea & Gulielmus de Papia.

2^a. (red) Preclarissimus liber elementorum Euclidis perspica-|cissimi: in artem Geometrie incipit quāfœlicissime: 137^b. COLOPHON: ℭ Opus elementorum euclidis megarensis in geometriam artem In id quoqꝫ | Campani perspicacissimi Commentationes finiunt. Impresuꝫ Vincentiæ per | Magistrum Leonarduꝫ de Basilea & Gulielmū de Papia Socios. Anno salutis | M.ccccłxxxxi. Vigesimo. Calen. Iuñ. Lector. Vale.

Folio. a^{10} b–r^8. 138 leaves, the first and last blank, without foliation. 4^a: 49 lines, 228 (with headline 238) × 120 (with diagrams 167) mm. Types: roman. Ornamental capitals. Three-side border on 2^a. Marginal diagrams.

A close reprint of Ratdolt's edition, omitting the dedication.

Hain *6694. Proctor 7130.

The partnership of these two printers produced only two books, of which this is the first.

Riccardi.

B.M. Bodl. C.T–S.

See pl. 3.

3.

1505, 25 October. Venice. Joannes Tacuinus.

1ᵃ. Euclidis megarēsis || philosophi platonicj || Mathematicaruꝫ disciplinarū Ianitoris: Habent in hoc volumi | ne quicūqꝫ ad mathematicā substantiā aspirāt: elemētorum libros. xiij. cū expositione | Theonis . . . quibus multa quae deerāt exlectiōe graeca sumpta | addita sub nec nō plurima subuersa ⁊ prepostere: voluta in Cāpani īterp̄tatōe: | ordinata digesta ⁊ castigata sunt. Quibus etiā nonnulla ab illo venerando. | Socratico philosopho mirādo iudicio structa habent̃ adiūcta. Deputa-|tum scilicet Euclidi volumē. xiiij. cū expositiōe Hypsi. Alex. Itidēqꝫ | ⁊ Phaeno. Specu. ⁊ Perspe. cum expositione Theonis. ac miran-|dus ille liber Datoꝝ cum expositiōe Pappi Mechanici vna cū | Marini dialectici protheoria. Bar. Zāber. Vene. Interp̄te. | ¶ Cum gratīa ⁊ Priuilegio per decēnium. [DEVICE.] 239ᵃ. ¶ Impressum Venetiis fœlicibus auibus opus est huiusmodi aureum & [ra]|rissimum: in ędibus Ioannis Tacuini librarii accuratissima dilligentia re|cognitum. Anno reconciliatæ diuinitatis. M.D.V. VIII. Klen-|das nouēbris Auspiciis fœlicissimæ Venetorum. Re. Pu. | Leornadi Lauretani: Principis sapientissimi: Cautū | est tamen dominii sanctione: nequis presens | opus Venetiis cudat: aut alibi impres|sum uēdere audeat: mulcta adiū|cta ut ī Priui. p̄ssius legit̃. || DEO [DEVICE] IMMORTALI || (beneath) GLORIA. 239ᵇ. Operis Canon.

Folio. $(1\text{–}4)^{10}$; A–Z AA–EE⁸ FF⁶. 240 leaves, the last blank. 13ᵃ: 42 lines (mixed types), 230 (with headline 240) × 134 (with diagrams 181) mm. Types: enunciations gothic, demonstrations roman; ll. 1, 2 of title printed from a wood block. Red printing on 11ᵃ. Ornamental capitals. Three-side (5-piece) border on 11ᵃ. Marginal diagrams.

1ᵃ, title; 1ᵇ, blank; 2ᵃ, B. Zamberti Protheoria; 7ᵇ, Vita Euclidis; 8ᵇ, Euclidis castigationes; 11ᵃ, Elementa, lib. I–XIV; 180ᵃ, Zambertus Paulo Pisano; 180ᵇ, Phaenomena; 191ᵇ, Specularia; 198ᵇ, B. Zambertus Ioanni Zamberto; 199ᵃ, Perspectiua; 210ᵃ, B. Zambertus Antonio Abiosio; 210ᵇ, Protheoria Marini; 213ᵃ, B. Zambertus Marino Georgio; 215ᵃ, Data; 239ᵃ, colophon and device; 239ᵇ, operis canon; 240, blank.

Riccardi.

B.M. C.T–S.

See pl. 4.

4.

1509, 22 May. Venice. Paganinus de Paganinis.

1ᵃ. (red) Euclidis || megarensis philo|(black)sophi acutissimi mathematicorumqꝫ omni|um sine controuersia principis opa a (red) Cam|pano (black) interprete fidissimo tralata (red) Q(black)ue cum | antea librariorum detestanda culpa mēdis | fedissimis adeo deformia eēnt: vt vix (red) Eu-|clidem (black) ipsum agnosceremus. (red) Lucas pacio|lus (black) theologus insignis: altissima (red) M(black)athe-|maticaꝝ disciplinarum scientia rarissimus | iudicio castigatissimo detersit: emendauit. | (red) F(black)iguras cētum ⁊ vndetriginta que in alijs | codicibus inuerse ⁊ deformate erant: ad re|ctam symmetriam concīnauit: ⁊ multas

❡EVCLIDIS MEGARENSIS CLARISSIMI philosophi Mathematicorumq; facile principis: primum ex Campano, deinde ex Theone Græco commentatore, interprete Bartholameo Zamberto Veneto, Geometrica Elementa. Liber duodecimus.

Eucli.ex Camp. Propositio prima.

OMnium duarum superficierum similium multiangularum inter duos circulos descriptarum est proportio alterius ad alteram: tanq̃ proportio quadratorum quæ ex diametris circulorum eas circũscribentium proueniunt.

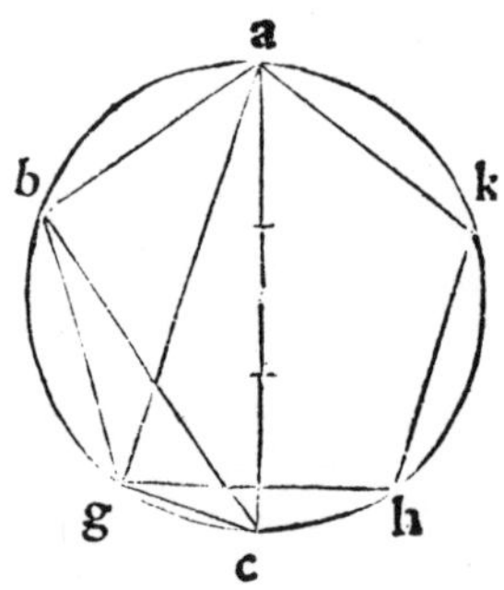

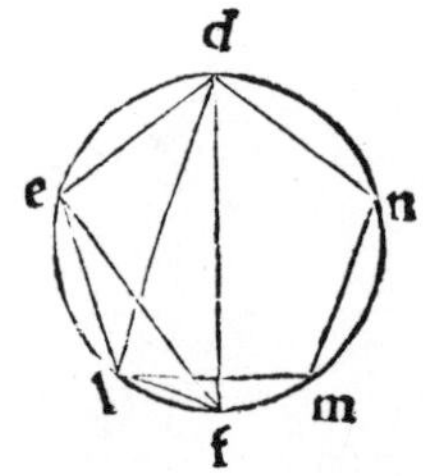

❡CAMPANVS. ❡Sint duo circuli a b c, d e f: quibus inscribãtur duæ q̃libet figuræ polygoniæ quæ ponãtur adinuicẽ similes. sintq; nũc/ pentagonæ inscriptæ vt docet 11 quarti: & ipsæ sint a b g h k, d e l m n. diametri quoq; circulorũ: sint a c & d f. Dico itaq; q̧ proportio pentagoni a b g h k ad pentagonũ d e l m n: est sicut quadratũ diametri a c ad quadratũ diametri d f. Protrahanẽ enĩ in vtroq; circulo duæ lineæ ab extremitate diametri ad extremitatẽ vnius lateris pentagoni diametro nõ cõterminalis: seinuicẽ cãcellãtes infra ipsum pentagonũ. in hoc quidẽ/ a g & c b: in illo autẽ/ d l & f e. Eritq; ex 6 sexti triangulus a b g: æquiãgulus triãgulo d e l. Nã cũ pentagoni ponãtur adinuicẽ similes: erũt ex diffinitione similiũ superficierũ angulus a b g æqualis angulo d e l: & latera ipsos continẽtia proportionalia/ videlicet proportio a b ad d e sicut b g ad e l. Cũ sint aũtex 20 tertij duo ãguli a c g & a g b sibi inuicẽ equales/ itẽq; duo alij d f e & d l e sibi inuicẽ equales: erũt duo qui sunt c & f adinuicẽ æquales ex hac cõmuni sciẽtia. quę æqualibus sunt æqualia: sibi quoq; æqua esse necesse est. Et quia ex prima parte 30 tertij vterq; duorum angulorũ a b c, d e f, est rectus: sequitur ex 32 primi duos triangulos a b c, d e f, esse equiangulos. Quare per 4 sexti proportio diametri a c ad diametrũ d f: est sicut lateris a b ad latus d e. Cũ itaq; ex secũda parte 18 sexti proportio duorum pentagonorũ est sicut proportio lateris a b ad latus d e proportio duplicata/ & per eandem proportio quadrati diametri a c ad quadratũ diametri d f sit sicut diametri a c ad diametrum d f duplicata: per hanc cõmunem scientiam quorum dimidia sunt æqualia ipsa quoq; adinuicem esse æqualia/ manifestum est quod propositum est.

Eucli.ex Zamb. Theorema primum. Propositio prima.

QVæ in circulis multangulæ figuræ: adinuicem se habẽt sicut quæ ex dimetientibus quadrata.

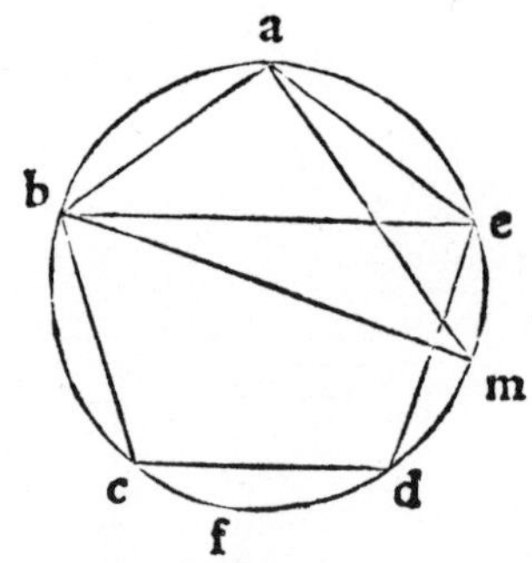

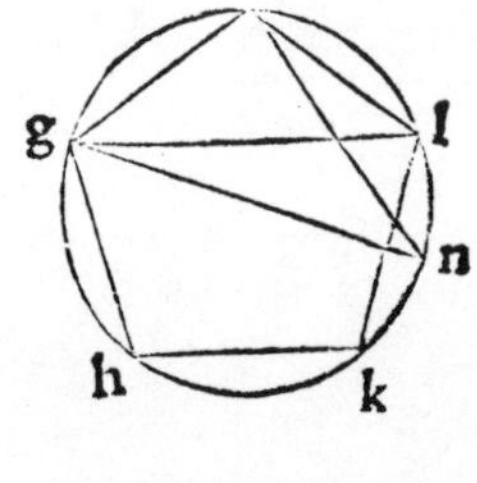

❡THEON ex Zamberto. ❡Sint circuli a b c d e, f g h k l: & in eis sint similes figurę multangulæ a b c d e & f g h k l. dimetientes autem circulorũ: sint b m, g n. Dico q̧ est sicut quadratũ quod ex b m ad id quod ex g n quadratũ: sic est multangulum a b c d e ad multangulum f g h k l. Connectantur enim b e, a m: g l, f n. Et quoniam multangulum a b c d e ipsi f g h k l multangulo simile est: æquus est & qui sub b a e angulus ei qui sub g f l, estq; sicut b a ad a e sic g f ad f l. Bina iam triangula sunt b a e & g f l: vnũ angulum vni angulo æquum habentia/ qui sub b a e ei qui sub g f l, circa autem æquos angulos latera proportionalia. ęquiangulum igitur est per primam diffinitionem sexti a b e triangulum: ipsi f l g triangulo. æqualis igitur est angulus qui sub a e b: ei qui sub f l g. Sed qui per 21 tertij sub a e b ei qui sub a m b est æqualis (in eandem namq; circũferẽtiam ierunt) qui autem sub f l g ei qui sub f n g: & qui sub a m b igitur ei qui sub f n g est ęqualis. Est autem & rectus qui sub b a m: ei qui sub g f n recto per 4 postulatum æqualis. reliquus igitur: reliquo est æqualis per 3 cõmunem sententiã. Aequiangulum igitur est triangulũ ne|cessarias addidit. (red) E(black)undē quoqȝ plurimis | locis intellectu difficilem cō-mentario|lis sane luculentis ⁊ eruditiss. ape|ruit : enarrauit : illustrauit (red) A(black)dhec | vt elimatior exiret (red) S(black)cipio ve|gius mediol. vir vtraqȝ | līgua : arte medica : subli|mioribusqȝ studijs | clarissimus diligē|tiam : ⁊ censurā | suā prestitit. || A. Paganius Paganinus Characteri-|bus elegantissimis accuratissi-|me imprimebat. 145ᵇ. ℂ Venetiis impressum per probum Virum Paganinum de Paganinis | de Brixia decreto tamen publico vt nullus ibidem totiꝗ dominio anno|rum. xv. curriculo imprimat aut imprimere faciat. Et alibi impressum | sub quouis colore in publicū ducat sub penis in dicto priuilegio contētis | Anno redemptionis nostre. M.D.VIIII. Klen. xi. Iunii. Leonardo | Lauretano Ve. Re. Pu. Gubernante. Pontificatus Iulii. II. Anno. VI.

Folio. a^{10} b–s^{8}. 146 leaves, the last blank, 2–145 numbered 2–144 with errors. 5^{a}: 57 lines (mixed types), 210 (with headline 219) × 94 (with diagrams 160) mm. Types: enunciations gothic, demonstrations roman; some greek. Ornamental capitals. Marginal diagrams.

1^{a}, title; 1^{b}, blank; 2^{a}, Lucas Paciolus de Burgo D. Francisco de Soderinis; 2^{b}, Danielis Caietani epigramma; 3^{a}, Danielis Caietani epistola Danieli Rainerio; 3^{b}, Franciscus Massarius Iacobo Cocco; 4^{a}, text; 145^{b}, colophon; 146, blank.

Riccardi.

B.M. Bodl. C.T–S.

See pls. 5, 6.

5.

1510, 26 March. Venice. Joannes Tacuinus.

A reissue of the sheets of the 1505 edition as far as the end of quire O. From this point the book has been reset in slightly different type, page for page and line for line.

COLOPHON: 239^{a}. ℂ Impressum Venetiis fœlicibus auibus opus est huiusmodi aureum & ra-|rissimum : in ædibus Ioannis Tacuini librarii accuratissima dilligētia re|cognitum. Anno reconciliatæ diuinitatis. M.D.X. VII. Klen|das Aprilis. Auspiciis fœlicissimæ Venetorum. Re. Pu. | Leonardi Lauredani : Prīcipis sapientissimi : Cautū | est tamen dominii sanctione : ne quis presens | opus Venetiis cudat : aut alibi impres|sum uēdere audeat : mulcta adiū|cta ut ī Priui. ꝑssius legiſ. || DEO [DEVICE] IMMORTALI || (beneath) GLORIA. 239^{b}. Operis Canon.

Device as in 1505 edition.

Some copies have a four-piece border on 11^{a} surrounding the text. See Ongania, *L'Arte della Stampa nel Rinascimento Italiano-Venezia*, pt. 2, p. 79.

The date of this edition has often been taken as 1517.

Riccardi.

B.M. C.T–S.

6.

[1516, after 7 January.] Paris. Henri Estienne.

1^{a}. CONTENTA. || EVCLIDIS Megarensis Geometricorum elemēto-|rum libri XV. | CAMPANI Galli trāsalpini in eosdem cōmentario|rum libri XV. | THEONIS Alexandrini Bartholamæo Zamberto | Veneto interprete, in tredecim priores, commentario-|rum libri XIII. || HYPSICLIS Alexādrini in duos posteriores, eodē | Bartholamæo Zamberto Veneto interprete, commē-|tariorum libri II. || VTCVNQVE NOSTER VALVIT LABOR | conciliata sunt hæc omnia, ad studiosorum non par-|uam (quam optamus) vtilitatem: id

Magnifico | D. FRANCISCO Briconneto postulāte. | Si hæc beneuole suscipiātur, & fructum | adferāt quē cupimus: alia eiusdē au⸗|thoris opera prodibūt in lucē, | successum præstāte deo, & | adiutoribus (vbivbi gē|tiū sint) ad bonarū | literarū institu|tionē pro⸗|be affe|ctis | Gallis, Italis, Germanis, Hispanis, Anglis. quibus | omnibus prospera imprecamur: & puram | pro dignitate veramq; co⸗|gnitionis lucem. || PARISIIS in officina Henrici Stephani e regione scho⸗|læ Decretorum. 261ª ℂ EVCLIDIS MEGARENSIS CLARISSIMI PHI|losophi Mathematicorūq; facile principis, primū ex Cāpa|ni, deinde ex Theonis in priores tredecim, & Hypsiclis | Alexandrini ī duos posteriores, Grȩcorū philosophorū tra|ditionibus, Bartholamæo Zāberto Veneto interprete: geo|metricorum Elementorum librorum quindecim. Finis. 261ᵇ [List of errata, followed by register.]

Folio. a–y⁸ z&⁶ A–H⁸ I¹⁰. 262 leaves, the last blank, 3–261 so numbered. 6ª: 55 lines (mixed types), 229 (with headline 234) x 109 (with diagrams 158) mm. Types: roman. Ornamental capitals. Marginal diagrams.

1ª, title; 1ᵇ, blank; 2ª, Francisco Briconneto Iacobus Faber; 3ª, text; 261ᵇ, errata, register; 262, blank.

Some copies have FINIA as the last word on 261ª and other unimportant variations of setting up on 261ᵇ.

The dedicatory letter of Jacobus Faber is dated 'postridie Epiphaniæ Domini M.D.XVI'. As the practice of the Paris printers was to date the new year from Easter, the year of printing may well be 1517.

Riccardi.

B.M. Bodl. C.T–S.

7.

1533, September. Basle. Johann Herwagen.

1ª. *ΕΥΚΛΕΙΔΟΥ* | *ΣΤΟΙΧΕΙΩΝ ΒΙΒΛ- ΙΕ⁻-* | *ΕΚ ΤΩΝ ΘΕΩΝΟΣ ΣΥΝ-*| *ΟΥΣΙΩΝ.* | *Εἰς τȣ αὐτȣ͂ τὸ πρῶτον, ἐξηγημάτων Πρόκλȣ βιβλ. δ⁻.* | Adiecta præfatiuncula in qua de disciplinis | Mathematicis nonnihil. || [DEVICE.] BASILEÆ APVD IOAN. HERVAGIVM ANNO | M.D.XXXIII. MENSE SEPTEMBRI. 198ª. BASILEÆ APVD IOAN. HERVAGIVM ANNO | M.D.XXXIII. MENSE SEPTEMBRI. 198ᵇ. [DEVICE.]

Folio. a⁶; *α–η⁶ θ⁴ ι–χ⁶ ψ⁴*; A–I⁶ K⁴. 198 leaves. Pagination after 6 preliminary leaves, 2–268, 1–115. Catchwords. 9ª: 47 lines (mixed types), 221 (with headline 228) x 131 (with marginalia 152) mm. Types: text greek, preliminary matter, colophon, &c., roman, marginalia italic. Ornamental capitals and headpieces. Border on 7ª. Diagrams in text.

1ª, title; 1ᵇ, blank; 2ª, Simon Grynæus Cutberto Tonstallo; 6ª, de methodo, et quonam pacto geometræ utantur, tum quomodo ad disciplinas cæteras sit transferenda, locus Galeni doctissime rem explicans; 7ª, *Στοιχεῖα*; 141ª, text of Proclus; 198ª, register, colophon; 198ᵇ, device.

The first edition of the original Greek text.

Riccardi.

B.M. Bodl. C.T–S.

8.

1536. Paris. Simon de Colines.

1ª. Orontij Finei Del|PHINATIS, REGII | Mathematicarum | professoris, | IN SEX PRIORES LIBROS | geometricorum elementorum | Euclidis Megarensis De⸗|monstrationes. || Quibus ipsius Euclidis textus græcus | suis lo⸗|cis insertus est: vnà cum interpretatione |

τὴν δυναμένην τὸ ἀπὸ τῆς ὅλης, καὶ τὸ ἀπὸ τοῦ ἐλάσσονος τμήματος. τούτων δὴ πάντων γνωρίμων ἡμῖν γενομένων, δῆλον ὅτι ἐὰν εἰς τὴν αὐτὴν σφαῖραν ἐγγραφῇ δωδεκάεδρον καὶ εἰκοσάεδρον, τὸ δωδεκάεδρον πρὸς τὸ εἰκοσάεδρον λόγον ἕξει ὃν εὐθείας οἵας δήποτε οὖν ἄκρον καὶ μέσον λόγον τετμημένης ἡ δυναμένη τὴν ὅλην, καὶ τὸ ἔλαττον τμῆμα πρὸς τὴν δυναμένην ὅλην καὶ ἔλαττον τμῆμα. ἐπεὶ γάρ ἐστιν ὡς τὸ δωδεκάεδρον πρὸς τὸ εἰκοσάεδρον, οὕτως ἡ τοῦ δωδεκαέδρου ἐπιφάνεια πρὸς τὴν τοῦ εἰκοσαέδρου, τουτέστιν ὡς ἡ τοῦ κύβου πλευρὰ πρὸς τὴν τοῦ εἰκοσαέδρου πλευρὰν, ὡς δὲ ἡ τοῦ κύβου πλευρὰ πρὸς τὴν τοῦ εἰκοσαέδρου, οὕτως ἐστὶν εὐθείας ἧς δήποτε οὖν ἄκρον καὶ μέσον λόγον τετμημένης ἡ δυναμένη τὴν ὅλην, καὶ τὸ μεῖζον τμῆμα πρὸς τὴν δυναμένην τὴν ὅλην καὶ τὸ ἔλαττον τμῆμα. ὡς ἄρα τὸ δωδεκάεδρον πρὸς τὸ εἰκοσάεδρον τῶν εἰς τὴν αὐτὴν σφαῖραν ἐγγραφομένων, οὕτως εὐθείας ἧς δήποτε οὖν ἄκρον καὶ μέσον λόγον τετμημένης. ἡ δυναμένη τὴν ὅλην καὶ τὸ μεῖζον τμῆμα πρὸς τὴν δυναμένην τὴν ὅλην καὶ τὸ ἔλαττον τμῆμα.

ΕΥΚΛΕΙΔΟΥ ΣΤΟΙΧΕΙΟΝ ΙΕ, ΚΑΙ ΣΤΕΡΕΩΝ ΠΕΜΠΤΟΝ, ΩΣ ΟΙΟΝΤΑΙ ΤΙΝΕΣ, ΩΣ ΑΛΛΟΙ ΔΕ ΥΨΙΚΛΕΟΥΣ Ἀλεξανδρέως, περὶ τῶν ε΄ σωμάτων δεύτερον.

α΄.

ΕΙΣ τὸν δοθέντα κύβον, πυραμίδα ἐγγράψαι. Ἔστω ὁ δοθεὶς κύβος ὁ αβγδεζηθ, εἰς ὃν δεῖ πυραμίδα ἐγγράψαι. ἐπεζεύχθωσαν αἱ αγ αε γε αθ εθ θγ. φανερὸν δὴ ὅτι τὰ αεγ αθε αθγ γθε τρίγωνα ἰσόπλευρά ἐστι. τετραγώνων γάρ εἰσι διάμετροι αἱ πλευραί. πυραμὶς ἄρα ἐστὶν ἡ αεγθ, καὶ ἐγγέγραπται εἰς τὸν δοθέντα κύβον.

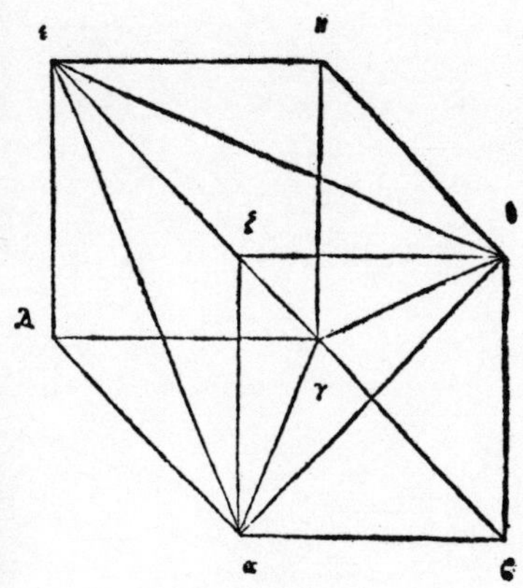

β΄.

ΕΙΣ τὴν δοθεῖσαν πυραμίδα ὀκτάεδρον ἐγγράψαι. Ἔστω ἡ δοθεῖσα πυραμὶς ἡ αβγδ, καὶ τετμήσθω δίχα τοῖς ε ζ η θ κ λ σημείοις, καὶ ἐπεζεύχθωσαν αἱ θη θλ εζ ζη, καὶ αἱ λοιπαί. καὶ ἐπεὶ ἡ αβ διπλῆ ἐστιν ἑκατέρας τῶν θη ηζ, ἴση ἄρα ἐστὶν ἡ θη τῇ ηζ, καὶ παράλληλος. ὁμοίως καὶ ἡ θη τῇ ζκ ἴση τέ ἐστι καὶ παράλληλος. ἰσόπλευρον ἄρα ἐστὶ τὸ θκζη, λέγω ὅτι καὶ ὀρθογώνιον. ἐὰν γὰρ ἀπὸ τῆς κλ κάθετοι ἀχθῶσιν ἐπὶ τὰ ἐπίπεδα τὰ εζβη ζγεη εζθη θκζη, ὁμοίως δείξομεν τὰ ἐπὶ τοῦ θκζη τετραγώνου ἰσόπλευρα.

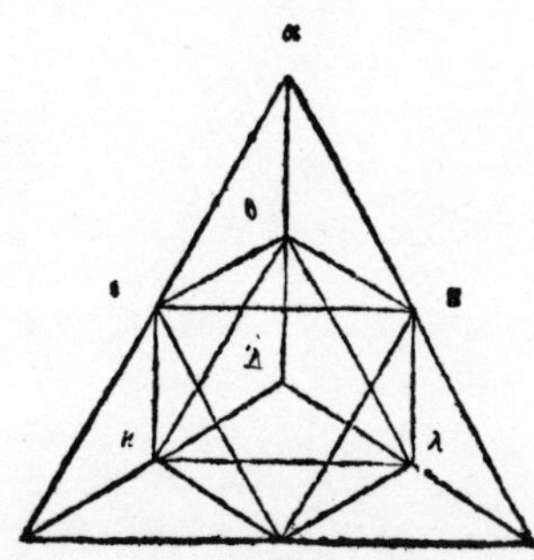

γ΄.

ΕΙΣ τὸν δοθέντα κύβον ὀκτάεδρον ἐγγράψαι. Ἔστω ὁ δοθεὶς κύβος ὁ αβγδεζηθ, καὶ εἰλήφθω τὰ κέντρα ἐφεστώτων τετραγώνων τὰ ψ ζ κ λ μ ν

CVM PRIVILEGIO | Regis ad decennium, || PARISIIS. | Apud Simonem Colinæum. | 1536. || Virescit vulnere virtus. 91ᵇ. SEXTI LIBRI GEOMETRICO⸗|rum Elementorū Euclidis Megarensis, Ex | Orontij Finei Delphinatis, Regij | Mathematicarum pro⸗|fessoris, tradi⸗|tione, || FINIS. || Virescit vulnere virtus. || [List of errata, register.]

Folio. ✱⁴; a–o⁶ p⁴. 92 leaves, the last blank. Pagination, after 4 preliminary leaves, 2–174. 8ᵃ: 45 lines (mixed types), 225 (with headline 234) x 140 (with marginalia 161) mm. Types: enunciations greek and roman, demonstrations roman; some italic. Ornamental capitals and headpieces. Elaborate border surrounding title on 1ᵃ. Diagrams in text.

1ᵃ, title; 1ᵇ, blank; 2ᵃ, Dedication to Francis I; 3ᵃ, ad lectorem; 4ᵃ, list of Finaeus' works; 4ᵇ, privilege; 5ᵃ, text; 91ᵇ, colophon, list of errata, register; 92, blank.

2nd edition, 1544, 3rd edition, 1551.

Riccardi.

B.M. C.T–S.

9.

1537, August. Basle. Johann Herwagen.

1ᵃ. EVCLIDIS MEGARENSIS MA⸗|THEMATICI CLARISSIMI ELEMEN⸗| torum Geometricorum. Lib. XV. || Cum expositione THEONIS in priores XIII à Bartholomæo Veneto latinitate do⸗|nata, CAMPANI in omnes, *&* HYPSICLIS Alexandrini in duos po⸗|stremos. || His adiecta sunt Phænomena, Catoptrica *&* Optica, deinde Protheoria Marini *&* Data, || Postremûm uero, Opusculum de Leui *&* Ponderoso, hactenus non nisum, eiusdem | autoris. [DEVICE.] BASILEÆ APVD IOHANNEM HERVAGIVM, | MENSE AVGVSTO. ANNO | M.D.XXXVII. || Cum priuilegio Cæsareo. 298ᵃ. BASILEÆ APVD IOHANNEM | HERVAGIVM, ANNO | M.D.XXXVII. | MENSE AVGV⸗|STO. 298ᵇ. [DEVICE.]

Folio. †⁴; a–z A–Z Aa–Cc⁶. 298 leaves. Pagination, after 4 preliminary leaves, 1–587. Catchwords. 12ᵃ: 56 lines (mixed types), 225 (with headline 229) x 133 (with marginalia 145) mm. Types: roman and italic; some greek. Ornamental capitals. Diagrams in text.

1ᵃ, title; 1ᵇ, Heruagius lectori; 2ᵃ, Philippus Melancthon studiosis adolescentibus . . . Vuittēberge. Mense Augusto 1537; 5ᵃ, Elementorum lib. i–xv; 246ᵃ, Zambertus Paulo Pisano; 246ᵇ, Phaenomena; 256ᵃ, B. Zambertus Lodouico Mocenico; 256ᵇ, Specularia; 262ᵇ, B. Zambertus fratri, Optica (Perspectiua); 273ᵃ, B. Zambertus Antonio Abiosio, Protheoria Marini; 275ᵃ, B. Zambertus Marino Georgio; 276ᵃ, Data; 297ᵃ, de leui & ponderoso fragmentum; 297ᵇ, additions of Campanus to Bk. iv; 298ᵇ, register colophon; 298ᵇ, device.

Reprinted in 1546, 1557 (Graesse), 1558.

In many copies the preface has been removed by the censor.

Riccardi.

B.M. C.T–S.

10.

1544. Paris. Simon de Colines.

1ᵃ. ❦ORONTII❦ | FINÆI DELPHINATIS, | REGII MATHEMATI⸗|CARVM LVTETIÆ | PROFESSORIS, || In sex priores libros Geometricorum | elementorum Euclidis Megarēsis demonstratio⸗|nes, Recēns auctæ, & emendatæ: vnà cum ipsius | Euclidis textu græco, & interpretatione | latina Bartholamæi Zamberti Ve⸗|neti. Omnia ad fidem geome-|tricam, per eundē Oron-|tium recognita. ❦ ‖ LVTETIÆ PARISIORVM, | Apud Simonem Colinæum. ‖ 1544. | Cum priuilegio Regis. ‖ Virescit vulnere virtus. 80ᵇ. ❧ SEXTI LIBRI GEOMETRICORVM | Elementorum Euclidis Megarensis, Ex Orontij | Finæi Delphinatis, Regij Mathematica-|rum Lutetiæ professoris, recèns | aucta & emendata | traditione, ‖ FINIS. ‖ Virescit vulnere virtus. ❦ ‖ Registrum.

Folio. ❧^{4}; a–d E–M^6 N^4. 80 leaves. Pagination, after 4 preliminary leaves, 1–152. 8ᵃ: 45 lines (mixed types), 231 (with headline 239) × 140 (with marginalia 167) mm. Types: enunciations greek and roman, demonstrations roman. Some ornamental capitals and headpieces. Border surrounding title on 1ᵃ. Diagrams in text.

A reprint of the 1536 edition with the dedicatory letter, preface, and bibliography brought up to date. Space is saved by the use of a smaller type for the demonstrations.

Riccardi.

B.M. Bodl. C.T–S.

See pl. 7.

11.

1546, August. Basle. Johann Herwagen.

1ᵃ. ♣ Euclidis Megarensis mathematici | clarissimi Elementorum geo-|metricorum libri xv. ‖ Cum expositione Theonis in priores XIII à Bartholomæo | Veneto Latinitate donata, Campani in omnes, & | Hysiclis Alexandrini in duos postremos. ‖ His adiecta sunt Phænomena, Catoptrica & Optica, | deinde Protheoria Marini & Data, ‖ Postremùm uerò, Opusculum de Leui & Ponderoso, | hactenus non uisum, eiusdem autoris. ‖ [DEVICE.] ‖ BASILEÆ, PER IOHANNEM HERVAGIVM, | MENSE AVGVSTO, ANNO | M.D.XLVI. ‖ Cum priuilegio Cæsareo. 298ᵃ. BASILEÆ, PER IOHANNEM | HERVAGIVM, ANNO | M.D.XLVI. | MENSE AV-|GVSTO. 298ᵇ. [DEVICE.]

Folio. †^{4}; a–z Aa–Zz AA–CC^6. 298 leaves. Pagination, after 4 preliminary leaves, 2–587. 10ᵃ: 55 lines (mixed types), 232 (with headline 237) × 132 mm. Types: roman and italic; some greek. Ornamental capitals. Headpieces. Diagrams in text.

Riccardi.

B.M. Bodl. C.T–S.

12.

1550, September. Basle. Johann Herwagen.

1ᵃ. EVCLIDIS | Megarensis, Philosophi & Mathe|MATICI EXCELLENTISSIMI, SEX LIBRI PRIORES, DE | Geometricis principijs, Græci & Latini, unà cum demonstrationibus | propositionum, absq; literarum notis, ueris ac proprijs, & alijs quibus-|dam, usum earum concernentibus, non citra maximum | huius artis studiosorum emolumen-|tum adiectis. ‖ ALGEBRÆ PORRO REGVLÆ, PROPTER NVME-|rorum exempla, passim propositionibus adiecta, his libris præmissæ | sunt, eædemq; demonstratæ. ‖ AVTHORE IOANNE SCHEVBELIO, IN | inclyta Academia Tubingensi Euclidis | professore ordinario. | [DEVICE.] | Cum gratia & priuilegio Cæsario, | ad quinquennium. ‖ BASILEÆ, PER IOAN-|nem Heruagium. 162ᵃ. BASILEÆ, PER IOANNEM | Heruagium, Anno salutis humanæ M.D.L. | Mense Septembri. 162ᵇ. [DEVICE.]

Folio. a^4; A–Z Aa–Pp^4 Qq^6. 162 leaves. Pagination, after 4 preliminary leaves, 2–315. Catchwords. Types: enunciations greek and roman, demonstrations roman. Ornamental capitals. Diagrams in text.

EVCLIDIS MEGARENSIS GRAECI PHILOSOPHI, ARITHMETICORVM *elementorum, Liber octauus.*

Ex Campano. Diffinitiones.

LAtera numerorum dicuntur, quorum multiplicatione numeri producuntur. 2 Superficialis appellatur numerus, qui sub duobus lateribus continetur. 3 Solidus uerò qui sub tribus, ex quorum continua multiplicatione habet procreari. 4 Quadratus, est numerus superficialis æqualibus lateribus consistens. 5 Cubus, est solidus ęqualibus consistens lateribus. 6 Similes dicuntur numeri superficiales siue solidi, quorũ latera sunt proportionalia.

Euclid. ex Camp. Propositio 1.

SI numerornm quotlibet continuæ proportionalitatis duo extremi fuerint contra se primi, eos omnes sęcundum suam proportionem minimos esse necesse est. 1

CAMPANVS. Sint continuè proportionales a, b, c, duoq; extremi qui sunt a, c, sint contra se primi, dico quòd in eadem proportione, non reperientur totidẽ minores. Si autem contingit, sint d, e, f, eritq; per 15 septimi, a ad c, sicut d ad f, & quia a & c sunt minimi in sua proportione per 23 eiusdem, sequitur per 21 ut a numeret d, & c, f, maiores scilicet minores, quod esse non potest.

a . . . . b . . . . . . c . . . .

d . . . e . . . . . f . . . .

Euclid. ex Zamb. *Theorema 1.* Propositio 1.

SI fuerint quotcunq; numeri continuè proportionales, extremi uerò ipsorum primi adinuicẽ fuerint, minimi sunt eandem rationem habentium eis. 2

THEON ex Zamb. *Sint quotcunq; numeri continuè proportionales α, β, γ, δ, extremi autem ipsorum hoc est α, δ, primi sunt adinuicem.* Dico quòd ipsi *α, β, γ, δ, minimi sunt eandem rationem habentium eis. Si autem non: sint minores ipsis α, β, γ, δ, ipsi ε, ζ, η, θ, in eadem ratione existentes eis.* Et quoniam *ipsi α, β, γ, δ, in eadẽ sunt ratione ipsis ε, ζ, η, θ, & æqualis est multitudo ipsorum ε, ζ, η, θ, at α, δ, multitudini ipsorum α, β, γ, δ, æquè igitur est sicut α ad δ, sic ε, ad θ, primi sunt adinuicem, primi uerò & minimi* (per 14 *septimi*) *minimi autem numeri metiuntur eandem rationem habentes æqualiter, antecedens antecedentem, & sequens sequentem* (per 21 *septimi.*) Metitur *igitur α, ipsum ε, maior minorem, quod* est *impossibile.* Igitur *ipsi ε, ζ, η, θ, minores existentes ipsis α, β, γ, δ, in eadem non sunt ratione ipsis.* Igitur *α, β, γ, δ, minimi sunt eandem rationem habentium eis:* quod oportebat demonstrare.

α . . . .

β . . . .

γ . . . .

δ . . . .

ε . . . . . .

ζ . . . .

η . . . .

θ . . . .

Euclid. ex Camp. Propositio 2.

NVmeros quotlibet continuæ proportionalitatis, secundum proportionem datam, minimos inuenire. 2

EVCLIDIS MEGARENSIS CLARISSIMI PHILOSOPHI, MATHEMAticorum facilè principis, primùm ex Campano, deinde ex Theone Græco cõmentatore, interprete Bartholomæo Zamberto Veneto, Geometricorũ elementorum Liber primus.

Ex Campano, triplex principiorum genus.

Primùm, Diffinitiones.

PVnctus est, cuius pars non est. 2 Linea, est lõgitudo sine latitudine: 3 cuius quidẽ extremitates, sunt duo pũcta. 4 Linea recta, est ab uno puncto ad aliũ breuissima extensio, in extremitates suas eã recipiens. 5 Superficies est, quæ longitudinem & latitudinem tantùm habet: 6 Cuius quidem termini, sunt lineę. 7 Superficies plana, est ab una linea ad aliam breuissima extensio, in extremitates suas eam recipiens.

Punctus, aut signum. Linea. Su per fi ci es.

8 Angulus planus, est duarũ linearũ alternus contactus, quarũ expãsio est super superficiẽ, applicatioq; nõ directa. 9 Quãdo autem angulũ continent duæ lineæ rectę, rectilineus angulus nominatur. 10 Quãdo recta linea super rectã steterit, duoq; anguli utrobiq; fuerint ęquales, eorũ uterq; rectus erit, lineaq; lineæ superstãs, ei cui superstat, perpẽdicularis uocat̃. 11 Angulus uerò qui recto maior est, obtusus dicitur. 12 Angulus uerò minor recto, acutus appellatur.

Angulus planus. Rectilineus. Angulus obliquus. a Acutus. b Obtusus. Linea perpend. c Rectus.

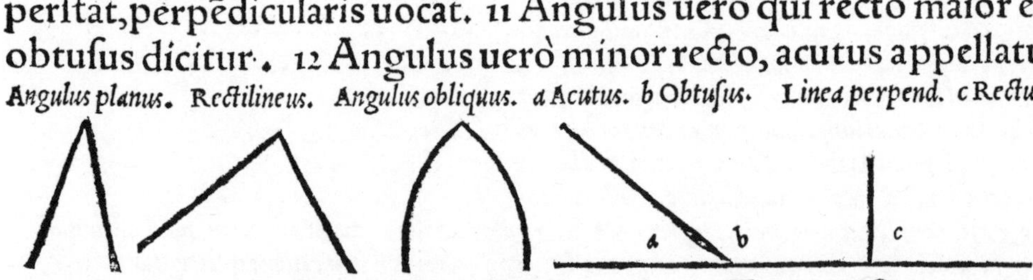

13 Terminus est, q̃d uniuscuiusq; finis est. 14 Figura est, quæ termino uel terminis continetur. 15 Circulus, est figura plana, una quidẽ linea cõtenta, quę circũferentia nominatur: in cuius medio punctus est, à quo omnes lineę rectę & ad circũferentiã exeũtes, sibi inuicem

1ᵃ, title; 1ᵇ, epigrams of M. Garbicius, Ioan. Sambucus; 2ᵃ, Ioannes Scheubelius, epistola nuncupatoria; 4ᵇ, de Euclide ex Proclo; 5ᵃ, Breuis regularum algebræ descriptio; 43ᵃ, Elementa, lib. 1–6; 162ᵃ, Scheubelius lectori, colophon; 162ᵇ, device.

Riccardi.

B.M. Bodl. C.T–S.

13.

1551. Paris. Regnault Chaudière.

Orontij Finæi Delphi-|NATIS, REGII MATHEMA|ticarum Lutetiæ professoris, || In sex priores libros Geometricorum | elementorum Euclidis Megarensis demonstrationes, recens | auctæ, & emendatæ: vnà cum ipsius Euclidis textu | Græco, & interpretatione Latina Bartholomęi | Zamberti Veneti. Omnia ad fidem geo-|metricam, per eundem Oron-|tium recognita. || Aeditio tertia || [DEVICE.] || LVTETIÆ PARISIORVM. || Apud Reginaldum Calderium. || 1551. || Virescit uulnere uirtus.

Quarto. ⁴; A–Z & Aa–Dd⁴. 116 leaves, 5–114 numbered 1–110. Catchwords. 7ᵃ: 42 lines (mixed types), 182 (with headline 185) x 102 (with marginalia 126) mm. Types: text greek, roman, italic, title roman, preliminary matter roman, italic. Ornamental capitals. Diagrams in text.

1ᵃ, title; 1ᵇ, blank; 2ᵃ, dedication to Francis I; 3ᵇ, ad Lectorem; 4ᵃ, epigrams; 5ᵃ, text; 115ᵃ, index operum ab O. Finæo conscriptorum; 116ᵇ, blank.

The dedicatory letters and bibliography are dated mensi octobri 1551.

Riccardi (who gives S. de Colines as printer).

Paris, Bib. Nat.

14.

1557. Lyons. Jean de Tournes & Guillaume Gazeau.

1ᵃ. IACOBI | PELETARII | CENOMANI, * || In Euclidis Elementa Geometrica | Demonstrationum | Libri sex. || Ad Carolum Lotharingum, Principem, | Cardinalemq; amplissimum. || LVGDVNI, | APVD IOAN. TORNÆSIVM | ET GVL. GAZEIVM. | M.D.LVII. || Cum Priuilegio Regis.

Folio. A⁴; a–o⁶; p⁶. 94 leaves. Pagination, after 4 preliminary leaves, 1–166, followed by 7 unnumbered leaves. 6ᵃ: 47 lines, 254 (with headline 262) x 143 mm. Types; roman and italic. Ornamental capitals. Headpieces. Elaborate border surrounding title. Diagrams in text.

1ᵃ, title; 1ᵇ, priuilegij summa; 2ᵃ, quæ ad sex libros Elementorum contulimus præcipua; 4ᵃ, Iacobi Peletarii præfatio; 5ᵃ, text; 88ᵃ, errata; 88ᵇ, blank; 89ᵃ, letters of Peletarius to various persons.

Riccardi.

B.M. Bodl. C.T–S.

15.

1558, March. Basle. Johann Herwagen & Bernhard Brand.

EVCLIDIS MEGA-|rensis mathematici clarissimi | Elementorum geometrico-|rum libri xv. | Cum expositione Theonis in priores xiii à Bartholamæo | Zamberto Veneto latinitate donata, Campani in | omnes, & Hypsiclis Alexandrini in du-|os postremos. || His adiecta sunt Phænomena, Catoptrica & Optica, | deinde Protheoria Marini, & Data. ||

Postremùm uerò, Opusculum de Leui & Ponderoso, | hactenus non uisum, eiusdem autoris. || [DEVICE.] || Cum priuilegio Cæsareo. || BASILEÆ, PER IOANNEM HER-|uagium, & Bernhardum Brand, Anno | M.D.LVIII. 298a. BASILEÆ, PER IOANNEM HERVA-| gium & Bernardum Brand, Anno M.D.|LVIII. mense Martio. 298b. [DEVICE.]

Folio. †4; a–z Aa–Zz AA–CC6. 298 leaves. Pagination, after 4 preliminary leaves and unnumbered page, 2–587. Catchwords. 12a: 55 lines (mixed types), 231 (with headline 235) × 132 (with marginalia 146) mm. Types: roman and italic; some greek. Ornamental capitals. Diagrams in text.

Graesse records an edition of 1557.

Riccardi.

C.T–S.

16.

1566. Strasburg. Josias Rihel.

ANALYSEIS GEO-|metricæ sex librorum | Euclidis. || PRIMI ET QVINTI FACTAE A | CHRISTIANO HERLINO: RELIQVÆ VNA CVM COM=|mentariis, & Scholiis perbreuibus in eosdem sex libros Geo=|metricos: à Cunrado Dasypodio. || CVM Indice *Πορισμάτων καὶ λημμάτων*: item aliarum rerum atque uerborum | memorabilium. | PRO SCHOLA ARGENTINENSI. || [DEVICE.] | CVM Gratia & Priuilegio Cæsareo ad annos octo. || EXCVDEBAT IOSIAS RIHELIVS. | M.D.LXVI.

Folio. a4; A–Z AA BB4. 104 leaves, the last blank, 5–102 numbered I–XCVII Text in 2 columns. Catchwords. 14a: col. 1, 54 lines, type-page measurement, 228 (with headline 240) × 132 mm. Types: enunciations greek and roman, demonstrations roman, headlines, &c., italic. Some ornamental capitals. Headpieces. Diagrams in text.

1a, title; 1b, Ioan. Sturmius lectori; 2a, C. Dasypodius amplissimo Senatorum ordini Reipublicæ Scaphusianæ; 3b, blank; 4a, text; 90a, Commentaria rerum geometricarum & scholia perbreuia sex librorum geom. Euclidis; 102b, Index rerum & uerborum; 104, blank.

Riccardi.

B.M. C.T–S.

17.

1566. Paris. Jean Le Royer.

EVCLIDIS | MEGARENSIS MATHEMA-|TICI CLARISSIMI ELEMENTA GEO-|METRICA, LIBRIS XV. AD GERMANAM GEOMETRIÆ INTEL-|ligentiam è diuersis lapsibus temporis iniuria contra-|ctis restituta, adimpletis præter maiorum spem, | quæ hactenus deerant, solidorum re-|gularium conferentiis ac in-|scriptionibus. || His accessit decimus sextus liber, de solidorum regularium sibi inuicem | inscriptorum collationibus, tum etiam cœptum opusculum de Composi-|tis regularibus solidis planè peragendum. || AVTHORE | Francisco Flussate Candalla. | AD | Carolum IX. Christianissimum Galliarum Regem. || [DEVICE.] PARISIIS, | Apud Iohannem Royerium typographum Regium. | Ex iussu & priuilegio Regis. | 1566.

Folio. ã6 ẽ4; A6 B–Z Aa–Zz AA–CC4 DD6. 214 leaves, 11–214 numbered 1–204. Catchwords. 13a: 47 lines (mixed types), 270 (with headline 278) × 154 mm. Types: roman and italic. Ornamental capitals. Diagrams in text.

1a, title; 1b, blank; 2a, dedication to Charles IX; 3b, various epigrams; 6a, Candalla lectori; 8a, Præfatio; 11a, text; 212a, de mixtis ct compositis regularibus solidis; 214b, errata.

No. 19

BTCVXY: Est autem ex dicto coroll. præcedentis propos. polyedrum DNEOFPQR, ad polyedrum ASBTCVXY, in triplicata proportione diametri DF, ad diametrum AC. Igitur & sphæra KLM, ad sphæram ABC, erit in triplicata proportione diametri DF, ad diametrum AC. Ponatur ut sphæra KLM, ad sphæram ABC, ita sphæra DEF, ad aliam sphæram $Z\alpha\beta$. Habebit igitur & sphæra DEF, ad sphærã $Z\alpha\beta$, proportionem triplicatã diametri DF, ad diametrũ AC.

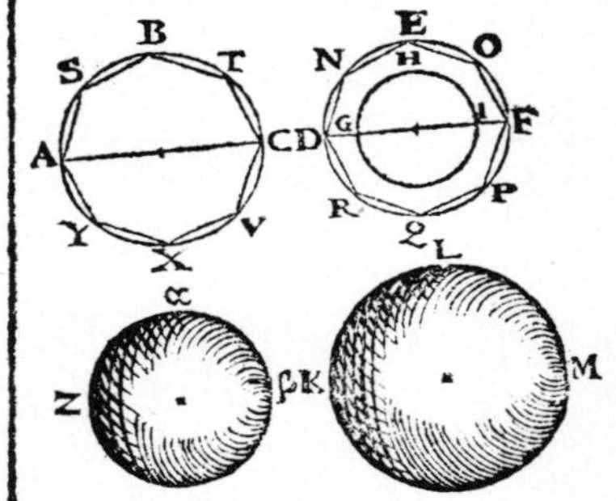

Et quia sphæra KLM, maior ponitur, quam sphæra DEF, erit quoq; sphæra ABC, maior, quam sphæra $Z\alpha\beta$. Quapropter sphæra DEF, ad sphæram $Z\alpha\beta$, minorem sphæra ABC, proportionem habet triplicatam diametri DF, ad diametrum AC. Quod est absurdum. Ostensum enim est nõ posse sphæram ad spheram alia sphera minorem, proportionem habere triplicatam diametrorũ. Non ergo habebit sphæra ABC, ad sphæram KLM, maiorem sphæra DEF, proportionem triplicatã diametri AC, ad diametrum DF; Sed neq; ad minorem habet, ut demonstratum est: Igitur habebit ad sphæram DEF, proportionem triplicatam diametri AC, ad diametrum DF. Sphæræ itaq; inter se sunt in triplicata ratione suarum diametrorum. Quod erat ostendendũ.

COROLLARIVM.

HINC fit, ita esse sphæram ad sphæram, ut polyedrum in illa descriptum ad polyedrum simile in hac descriptum. Quia tã sphæra ad sphæram, quam polyedrum ad polyedrum habet triplicatam diametrorum proportionem, ut demonstratum est.

FINIS ELEMENTI DVODECIMI.

EVCLIDIS

EVCLIDIS
ELEMENTVM XIII.
Et Solidorum tertium.

THEOR. I. PROPOS. I.

I.

SI recta linea secundum extremam & mediam rationem secetur; maius segmentum assumens dimidiam totius, quintuplũ potest eius, quod a dimidia totius describitur, quadrati.

SECETVR recta AB, in C, extrema ac media ratione, sitq; maius segmentũ AC. Producta autem BA, ad D, sit AD, dimidia totius AB. Dico quadratũ rectæ CD, quintuplum esse quadrati rectæ AD. Describatur.n. super CD, quadratũ CE, in quo ducta diametro DF, ducatur AG, ipsi DE, parallela, secans diametrũ in H, pũcto, p quod ducatur IK, parallela ipsi CD. Producta deinde GA, perficiatur quadratũ AM, protrahaturq; FC, ad N. eruntq; ex coroll. propos. 4. lib. 2. AI, KG, quadrata rectarũ AD, AC. Quia ergo est ut AB, ad AC, ita AC, ad CB; erit rectangulũ CM, sub AB, CB, æquale ipsi KG, quadrato rectæ AC. Deinde cũ ponatur AB, dupla ipsius AD, sitq; AL, ipsi AB; & AH ipsi AD, æqualis; erit quoq; AL, ipsius AH, dupla: Est autem ut

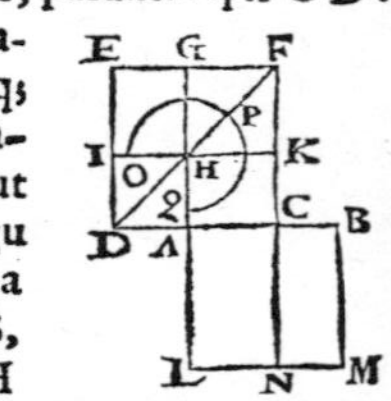

3. defin. sex.
17. sexti.

Reprinted in 1578.
Riccardi.
B.M. Bodl.

18.

1572. Pesaro. Camillo Francischini.

EVCLIDIS | ELEMENTORVM | LIBRI XV. | Vnà cum Scholijs antiquis. | A FEDERICO | COMMANDINO | VRBINATE | NVPER IN LATINVM | conuersi, commentarijsq́ue | quibusdam illustrati. || [in lower panel of border] PISAVRI, MDLXXII. | Cum Priuilegio Pont. Max. 267ᵇ. PISAVRI | CVM LICENTIA SVPERIORVM. || APVD CAMILLVM FRANCISCHINVM. | M.D.LXXII.

Folio. $*, **^6$; A–Z Aa–Zz Aaa–Sss^4. 268 leaves, the last blank, 13–267 numbered 1–255. Catchwords. 20^a: 53 lines (mixed types), 231 (with headline 235) × 134 (with marginalia) 150 mm. Types: roman and italic. Ornamental capitals. Elaborate architectural border signed IACOBVS Chriegher German' surrounding title on 1^a. Diagrams in text.

1^a, title; 1^b, blank; 2^a, Gregorii XIII priuilegium; 3^a, Commandinus Francisco Mariæ II Urbinatum Principi; 4^b, Commandini prolegomena; 8^b, index eorum, quæ in his libris demonstrantur præter ea, quę Euclidis sunt; 12^b, blank; 13^a, text; 267^b, colophon; 268, blank.

Riccardi.
B.M. Bodl. C.T–S.

19.

1574. Rome. Vincenzo Accolti.

Vol. 1. EVCLIDIS | ELEMENTORVM | LIBRI XV. | ACCESSIT XVI. DE SO-|lidorum Regularium com-|paratione. | OMNES PERSPICVIS DE-|monstrationibus, accuratisq́; scho-|lijs illustrati. || AVCTORE || CHRISTOPHORO CLAVIO | BAMBERGENSI. | SOCIETATIS IESV. || ROMÆ, | Apud Vincentium Accoltum. 1574. Vol. 2, 300^a. [DEVICE.] ROMÆ, | Apud Vincentium Accoltum. | M.D.LXXIIII. || Cum licentia Superiorum.

Octavo. 2 vol. Vol. 1: a–e^8; A–Z Aa–Ss^8 Tt^4. 372 leaves, the last blank, 41–371 numbered 1–331. Vol. 2: A–Z Aa–Oo^8 Pp^4. 300 leaves, 2–300 so numbered. Catchwords. Vol. 1, 11^a: 39 lines, 130 (with headline and rules 140) × 73 (with rules 89) mm. Types: roman and italic. Ornamental capitals. Rules round type-page. Borders (architectural) enclosing titles. Diagrams in text.

The title of vol. 2 agrees exactly with that of vol. 1 except that ll. 2, 3 read 'POSTERIORES | Libri sex a x. ad xv'.

Riccardi.
B.M. Bodl. C.T–S.

20.

1577. Leipzig. Johann Steinman.

Εὐκλείδȣ ϛοιχείων | *βίβλια ἕξ.* | EVCLIDIS | ELEMENTORVM GEO-|METRICORVM LIBRI SEX | conuersi in latinum sermonem à | Ioach. Camerario. || QVIBVS ADIECTÆ SVNT | trium priorum librorum Demonstrationes, atque | editæ in gratiam & vtilitatem studiosorum Ma-|thematices in Acad. Lips. à || Mauritio Steinmetz Gersb. | Medicinæ Licentiato. | [DEVICE.] | LIPSIÆ | Anno M.D LXXVII. 136^b. LIPSIÆ, | IMPRIMEBAT IOHANNES | STEINMAN. || Anno | M.D.LXXVII.

Octavo. ɔc[8], (?)[8]; A–P[8]. 136 leaves. Pagination, after 16 preliminary and 1 unnumbered page, 2–240. Catchwords. 9[a]: 32 lines, 121 (with headline 125) x 64 mm. Types: greek, roman, italic. Ornamental capital on 2[a]. Diagrams in text.

Riccardi.

C.T–S.

21.

1578. Paris. Iacques du Puys.

EVCLIDIS | MEGARENSIS MATHEMA-|TICI CLARISSIMI ELEMENTA, | LIBRIS XV. AD GERMANAM GEO-|metriæ intelligentiam è diuersis lapsibus | temporis iniuria contractis | restituta. | ❦ || Adimpletis præter maiorum spem, quæ hactenus deerant, | solidorum regularium conferentiis ac inscriptionibus. || Accessit decimussextus liber, de solidorum regularium sibi inuicem | inscriptorum collationibus. | Nouissimè collati sunt decimusseptimus & decimus octauus, priori | editione quodammodo polliciti, de componendorum, inscribendo-|rum & conferendorum compositorum solidorum inuentis, or-|dine & numero absoluti. || AVTHORE | D. Francisco Flussate Candalla. || AD | CAROLVM IX. CHRISTIANISSIMVM | GALLIARVM REGEM. || [DEVICE.] || LVTETIÆ, | APVD IACOBVM DV PVYS. | M. D. LXXVIII. 300[a]. LVGDVNI, | EX OFFICINA IOANNIS TORNÆSII, TYPOGRA-|PHI REGII. M.D. LXXVIII.

Folio. *[6]; **[6]; a–z aa–zz A B[6]. 300 leaves. Pagination, after 12 unnumbered leaves, 1–575. Catchwords. 19[a]: 51 lines (mixed types), 265 (with headline 272) x 141 mm. Types: roman; some italic. Ornamental capitals and headpieces. Diagrams in text.

1[a], title; 1[b], blank; 2[a], dedication to Charles IX; 4[a], epigrams; 7[a], Fr. Fluss. Candalla lectori; 9[b], Præfatio; 13[a], text.

Riccardi.

Bodl.

22.

1589. Rome. Bartolommeo Grassi.

EVCLIDIS | ELEMENTORVM LIB. XV || ACCESSIT XVI. DE SOLIDORV̄. | Regularium cuiuslibet intra quodlibet comparatione. || OMNES PERSPICVIS DEMON-| strationibus, accuratisq; scholijs illus-|trati: nunc iterum editi, ac multa℞ | rerum accessione locupletati: || AVCTORE | CHRISTOPHORO CLAVIO | BAMBERGENSI | È SOCIETATE IESV || ROMÆ APVD BARTHOLOMAEVM | Grassium: M D LXXXIX. | PERMISSV SVPERIORVM. Vol. 1, 468[a]. ROMÆ, Apud Sanctium, & Soc. | M.D.LXXXIX.

Octavo. 2 vol. Vol. 1: *[8]; A–Z Aa–Zz Aaa–Lll[8] Mmm[4]. 468 leaves. Pagination, after 8 preliminary leaves, 1–918, the last leaf unnumbered. Vol. 2: A–Z Aa–Zz Aaa–Ooo[8]; a–f[8]. 528 leaves. Pagination, after 1 unnumbered leaf, 3–959, followed by 48 unnumbered leaves. Vol. 1, 10[a]: 31 lines, 129 (with headline 135) x 92 mm. Types: roman and italic. Ornamental capitals. Type-page enclosed within rules. Engraved title-pages with border. Diagrams in text.

The title of vol. 2 agrees with that of vol. 1, except that POSTERIORES LIB. IX has been substituted in l. 2 for ELEMENTORVM LIB. XV.

Riccardi.

B.M.

No. 22

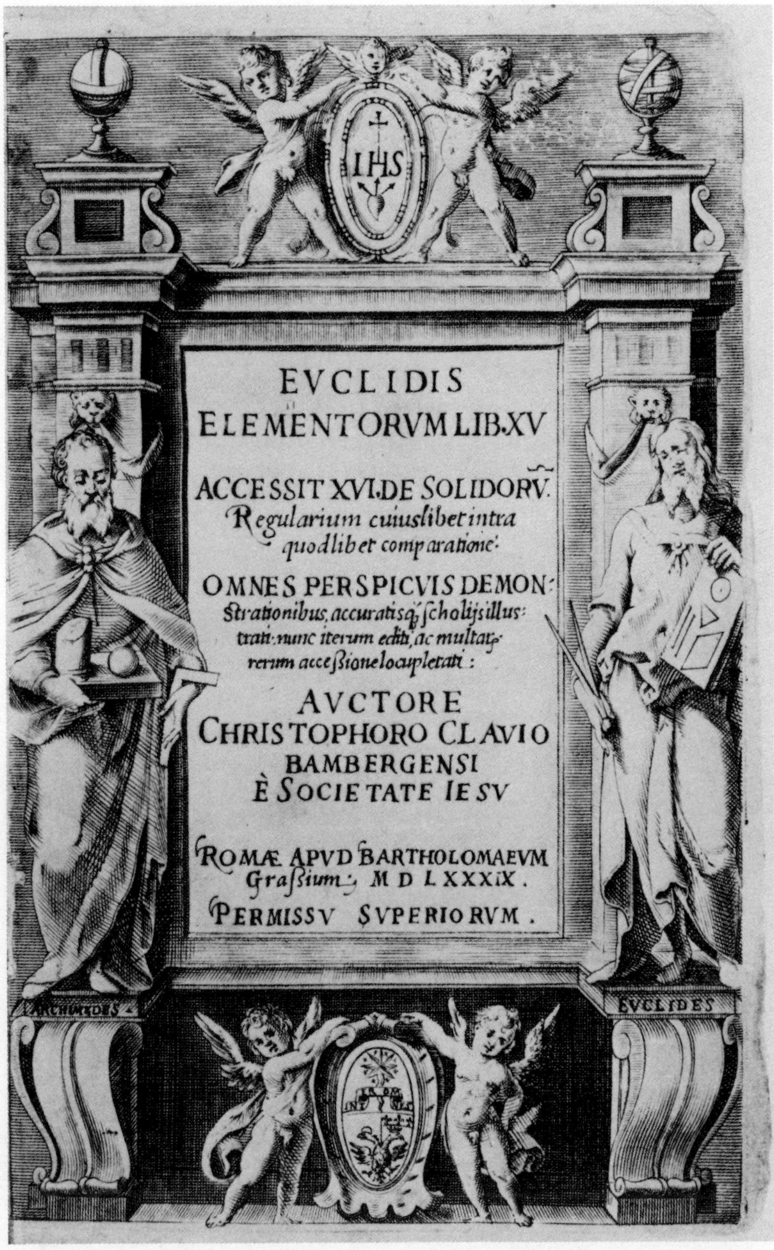
EVCLIDIS
ELEMENTORVM LIB. XV
ACCESSIT XVI. DE SOLIDORV̄.
Regularium cuiuslibet intra
quodlibet comparatione.
OMNES PERSPICVIS DEMON-
strationibus, accuratisq; scholijs illus-
trati: nunc iterum editi, ac multarum
rerum accessione locupletati:
AVCTORE
CHRISTOPHORO CLAVIO
BAMBERGENSI
È SOCIETATE IESV
ROMAE APVD BARTHOLOMAEVM
Grassium. M D L XXXIX.
PERMISSV SVPERIORVM.

23.

1591. Cologne. Joannes Baptista Ciotti.

EVCLIDIS | ELEMENTORVM | LIBRI XV. || Accessit XVI. de SOLIDORVM REGV-|LARIVM cuiuslibet intra quodli-|bet comparatione. || Omnes perspicuis DEMON-STRATIO-|NIBVS, accuratisque SCHOLIIS illustrati, | ac multarum rerum accessione | locupletati: || Nunc tertiò editi, summaq; diligentia recogniti, | atque emendati. || Auctore | CHRISTOPHORO CLAVIO BAM-|BERGENSI, è Societate IESV. || COLONIÆ, | Expensis IOH. BAPTISTAE CIOTTI | CIƆ IƆ XCI.

Folio. 2 vol. Vol. 1: ɔ:c⁸; a–z A–G⁶. 188 leaves. Pagination, after 8 preliminary leaves, 1–359. Vol. 2: A–Z Aa–Kk⁶. 198 leaves. Pagination, after 1 unnumbered leaf, 3–355, followed by 20 unnumbered leaves. Catchwords. Vol. 1, 12ª: 56 lines (mixed types), 248 (with headline 253) × 142 (with marginalia 155) mm. Types: roman and italic. Ornamental capitals. Elaborate border enclosing title. Diagrams in text.

The title-page of vol. 2 agrees exactly with that of vol. 1 except for the modification in l. 2 noted in preceding edition. The printer's device occurs on the blank last page of each volume. Ciotti, who was printing in Venice by 1592, is stated by F. Kapp, *Geschichte des deutschen Buchhandels bis in das siebzehnte Jahrhundert*, to have had a printing office in Cologne from 1587–1594. More probably he was a publisher with a branch at Cologne for the sale of books printed in Italy.

Riccardi.

B.M. Bodl. C.T–S.

24.

1598. Paris. Veuve G. Cavellat.

1ª. EVCLIDIS | ELEMENTORVM | LIBRI XV. GRAE-|cè & Latinè, || Quibus, cùm ad omnem Mathematicæ scientiæ | partem, tùm ad quamlibet Geometriæ tractatio-|nem, facilis comparatur aditus. || *Επίγραμμα παλαιόν.* | *Σχήματα πέντε Πλάτωνος, ἃ Πυθαγόρας σοφὸς* | *εὗρε.* | *Πυθαγόρας σοφὸς εὗρε, Πλάτων δ' ἀρίδηλ' ἐδί-*|*δαξεν,* | *Εὐκλείδης ἐπὶ τοῖσι κλέος περικαλλὲς ἔτευξεν.* | [DEVICE.] | PARISIIS, | Apud Viduam Guileilmi CAVELLAT, | sub Pellicano, monte D. Hilarij. || 1598. 176ᵇ. [DEVICE.]

Octavo. *A*B⁸; C–V⁸. 176 leaves, 2–175 numbered 3–150. Catchwords at ends of quires. 21ª: 27 lines (mixed types), 127 (with headline 132) × 75 mm. Types: text greek and roman, preface italic. Ornamental capital on 2ª. Headpieces. Diagrams in text with fleurons.

Riccardi.

Paris, Bibl. Nat.

25.

1600. Cologne. Goswin Cholinus.

EVCLIDIS | ELEMENTORVM | LIBRI XV. | QVIBVS, CVM AD OM-|NEM MATHEMATICÆ SCIEN=|TIÆ PARTEM, TVM AD QVAMLI-|bet Geometriæ tractationem | facilis comparatur | aditus. | [DEVICE.] | COLONIÆ | Apud Gosuinum Cholinum | M.D.C. | Cum Gratia *&* Priuilegio.

Octavo. A–O⁸ P⁴. 116 leaves. Pagination, after 15 preliminary leaves, 3–201, 204, 205, the last page blank. Catchwords. 9ª: 30 lines (italic), 125 (with headline 129) × 64 mm. Types: text, roman; preliminary matter, italic; some greek. Ornamental capital on 2ª. Diagrams with floral ornaments in text.

Riccardi.

C.T–S.

II

EDITIONS IN GREEK AND LATIN OF THE ENUNCIATIONS ONLY

26.

1545. Rome. Antonio Blado.

ΕΥ̓ΚΛΕΊΔΟΥ ΣΤΟΙΧΕΊΩΝ | *ΒΙΒΛΊΑ ιε΄.* || EVCLIDIS ELEMENTORVM | LIBRI XV. || [woodcut portrait of Euclid] || ROMÆ. MDXXXXV. || Cum priuilegio S. D. N. & Pauli III, *&* inclytæ | Reip. Venetiarum, ad quinquennium. 55^a. Romæ, apud Antonium bladum | Asulanum.

Octavo. *A–H*^{8}. 56 leaves, the last blank. 4^a: 25 lines, 117 (with book number 121) × 71 mm. Types: text greek, preface, portions of title, colophon italic, title roman capitals. Ornamental capital on 2^a.

The enunciations only without diagrams.

Riccardi.

B.M. Bodl. C.T–S.

See pl. 8.

27.

1545. Paris. Louis Grandin.

[Euclidis elementa mathematica.] 3^a. ❧ Euclidis elemēta | MATHEMATICA. || ℂ Diffinitio 1. . . . EVCLIDIS | Liber decimusquintus. || ℂ Propositio 1. | . . . FINIS. || PARISIIS. | Excudebat Ludouicus Grandinus. | 1545.

Octavo. a–f^8 g^4. 52 leaves. Pagination [2–4?] 5–104. 6^a: 27 lines, 122 (with headline 127) × 70 mm. Type: text italic. Ornamental capital.

The copy in the Bibliothèque Nationale, from which this description is taken, lacks the first two sheets of quire a.

Enunciations only translated by Ramus (Pierre de La Ramęe).

Riccardi.

Paris, B. N.

28.

1549. Leipzig. Valentin Bapst.

Εὐκλείδου στοιχείων βιβλία ἕξ. | EVCLIDIS | ELEMENTORVM GEO⸗|METRICORVM LIBRI SEX, | CONVERSI IN LATINVM | sermonem à Ioach. Camerario. | Edebat Lipsiæ Georg. Ioach. Rhet. | [device: an obelisk] | EXPRIMENTE VALENTINO | PAPA. ANNO M.D.XLIX.

Octavo. A^8; B–H^8 I^4. 68 leaves. Pagination, after 8 unnumbered leaves, 2–119. Catchwords. 11^a: 27 lines (mixed types), 121 (with headline 130) × 70 mm. Types: text, greek and roman, preface italic with some greek. Ornamental capital on 1^b. Diagrams in text.

Riccardi.

C.T–S.

See pl. 8.

29.

1549. Paris. Thomas Richard.

EVCLIDES. || [DEVICE.] || PARISIIS. || Apud Thomam Richardum, sub Bibliis | aureis, è regione Collegij Remensis.| 1549. 55ª. PARISIIS. | Excudebat Thomas Richardus. | 1549.

Octavo. a–g⁸. 56 leaves, the last blank, 2–55 so numbered. 6ª: 27 lines, 131 (with headline 136) × 71 mm. Types: preface roman, remainder italic. Ornamental capitals on 1ᵇ, 3ª.

The preface of the editor translator, Petrus Ramus (P. de La Ramée), is dated 5 cal. Feb. [28 Jan.] 1544.

Riccardi.

B.M.

30.

1557. Paris. Guillaume Cavellat.

❧ EVCLIDIS | ELEMENTORVM | LIBRI XV. GRÆ-|cè & Latiné, | Quibus, cùm ad omnem Mathematicæ scientiæ | partem, tùm ad quamlibet Geometriæ tra-|ctationem, facilis comparatur aditus. || *Επίγραμμα παλαιόν.* || *Σχήματα πέντε Πλάτωνος, ἅ Πυθαγόρας σο-|φὸς εὗρε.* | *Πυθαγόρας σοφὸς εὗρε, Πλάτων δ' ἀρίδηλ' ἐδί-|δαξεν,* | *Εὐκλείδης ἐπὶ τοῖσι κλέος περικαλλὲς ἔτευξεν.* | [DEVICE.] | LVTETIÆ, | Apud Gulielmum Cauellat, in pingui Gallina, | ex aduerso collegij Cameracensis. || 1557. 176ᵇ. [DEVICE.]

Octavo. ã ẽ⁸; A–V⁸. 176 leaves, 17–104 numbered 1–88, 105–176 numbered 59–130. Catchwords at ends of quires. 21ª: 28 lines, 128 (with headline 134) × 70 mm. Types: text greek and roman, title, preface italic. Ornamental capital and headpieces. Diagrams with fleurons.

The first edition of Gracilis with elaborately ornamented diagrams, reprinted at Paris in 1573, and closely imitated at Cologne in 1564, 1587.

Riccardi.

B.M.

Another issue, with date 1558.

Riccardi.

B.M. C.T–S.

31.

1564. Cologne. Maternus Cholinus.

EVCLIDIS | ELEMENTORVM LIBRI | XV. GRÆCE ET LATINE, | Quibus, cùm ad omnem Mathematicæ scientiæ | partem, tùm ad quamlibet Geometriæ tra=| ctationem, facilis comparatur aditus. || *Επίγραμμα παλαιόν.* | *Σχήματα πέντε Πλάτωνος, ἅ Πυθαγόρας σοφὸς* | *εὗρε.* | *Πυθαγόρας σοφὸς εὗρε, Πλάτων δ' ἀρίδηλ' ἐδίδαξεν,* | *Εὐκλείδης ἐπὶ τοῖσι κλέος περικαλλὲς ἔτευξεν.* || [DEVICE.] || COLONIÆ, | Apud Maternum Cholinum. | M. D. LXIIII.

EVCLIDIS ELEMENTORVM

LIBRI XV. GRAE-cè & Latiné,

Quibus, cùm ad omnem Mathematicæ ſcientiæ partem, tùm ad quamlibet Geometriæ tractationem, facilis comparatur aditus.

Ἐπίγραμμα παλαιόν.

Σχήματα πέντε Πλάτωνος, ἃ Πυθαγόρας σο-
φὸς εὗρε.
Πυθαγόρας σοφὸς εὗρε, Πλάτων δ' ἀρίδηλ' ἐδί-
δαξεν,
Εὐκλείδης ἐπὶ τοῖσι κλέος περικαλλὲς ἔτευξεν.

LVTETIAE,
Apud Gulielmum Cauellat, in pingui Gallina, ex aduerſo collegij Cameracenſis.
1557.

cuius eſt ſegmentum.

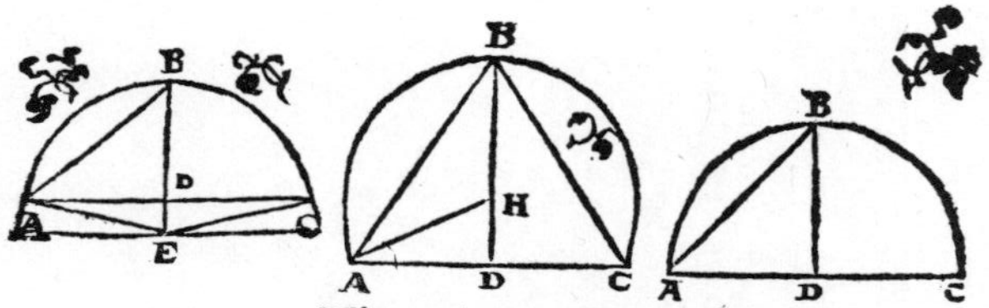

Theorema 23. Propoſitio 26.

In æqualibus circulis, æquales anguli æqualib. peripherijs inſiſtunt ſiue ad centra, ſiue ad peripherias conſiſtituti inſiſtant.

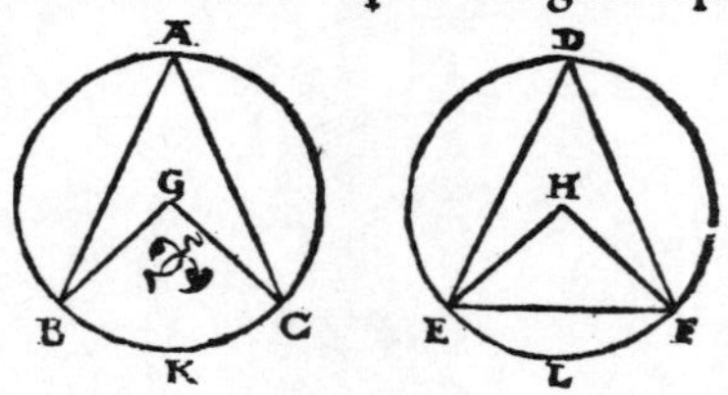

Theorema 24. Propoſitio 27.

In æqualibus circulis, anguli qui æqualibus peripherijs inſiſtũt, ſunt inter ſe æquales ſiue ad centra, ſiue ad peripherias conſtituti inſiſtant.

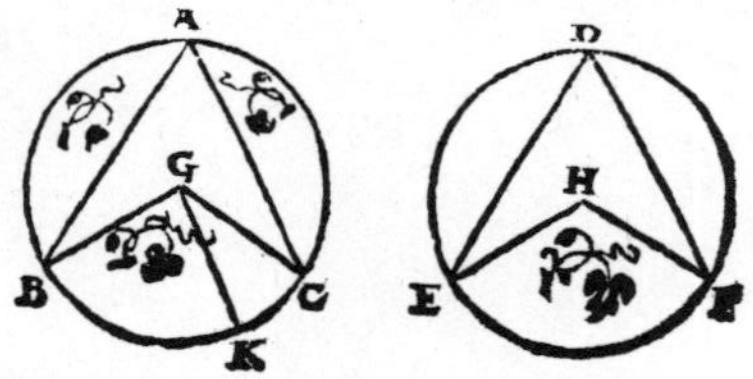

Theore-

Theorema 25. Propoſitio 28.

In æqualibus circulis æquales rectæ lineæ æquales peripherias auferũt, majoré quidem majori, minorem autem minori.

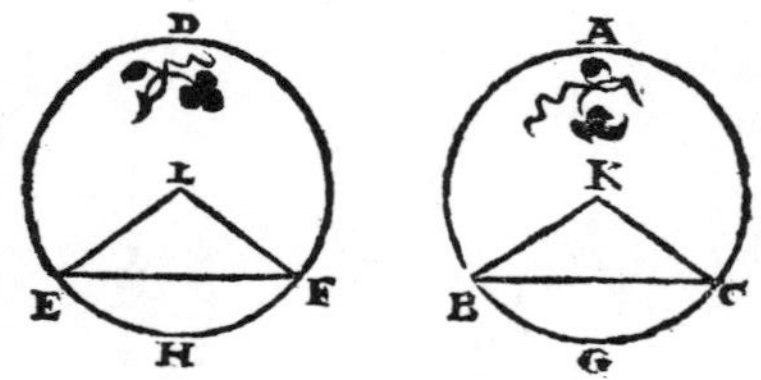

Theorema 26. Propoſitio 29.

In ęqualibus circulis, æquales peripherias æquales rectæ lineæ ſubtendunt.

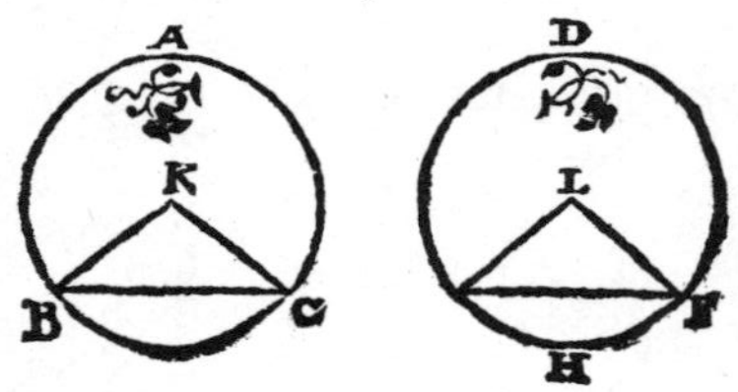

Problema 4. Propoſitio 30.

Datam peripheriam biſariam ſecare.

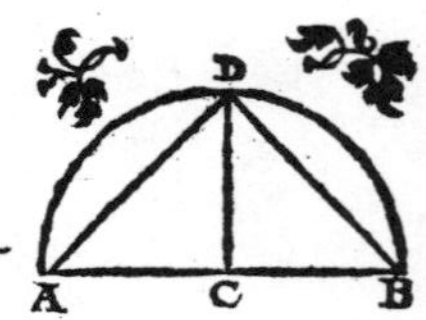

Theorema 27. Propoſitio 31

In circulo angulus qui in ſemicirculo, rectus

Octavo. *αβ*⁸; A–V⁸. 176 leaves, 16 and 176 blank, 17–175 numbered 1–159. Catchwords. 3ᵃ: 26 lines, 113 (with headline 119) × 67 mm. Types: text greek and roman, title, preface italic. Ornamental capital and headpieces. Diagrams.

A close reprint and imitation of Cavellat's 1557 Paris edition.

Riccardi.

B.M.

32.

1573. Paris. Hiérosme de Marnef & Guillaume Cavellat.

EVCLIDIS | ELEMENTORVM | LIBRI XV. GRÆ-|cè & Latinè, | Quibus, cùm ad omnem Mathematicæ scientiæ | partem, tùm ad quamlibet Geometriæ tractatio-|nem, facilis comparatur aditus. || *Ἐπίγραμμα παλαιόν.* | *Σχήματα πέντε Πλάτωνος, ἃ Πυθαγόρας σοφὸς* | *εὗρε.* | *Πυθαγόρας σοφὸς εὗρε, Πλάτων δ' ἀρίδηλ' ἐδί-|δαξεν,* | *Εὐκλείδης ἐπὶ τοῖσι κλέος περικαλλὲς ἔτευξεν.* | [DEVICE.] | PARISIIS, | Apud Hieronymum de Marnef, *&* Gulielmum | Cauellat, sub Pellicano, monte D. Hilarij. | 1573. 176ᵇ. [DEVICE.]

Octavo. *AB*⁸; C–Y⁸. 176 leaves. Pagination, after unnumbered title leaf, 3–350, followed by an unnumbered leaf. Catchwords at ends of quires. 4ᵃ: 26 lines, 124 (with headline 128) × 74 mm. Types: text greek and roman, title, preface italic. Ornamental capital and headpieces. Diagrams.

A reprint of the 1557 edition with corrections.

Riccardi.

B.M. Bodl.

33.

1587. Cologne. Maternus Cholinus.

EVCLIDIS | ELEMENTORVM | LIBRI XV. || Quibus, cùm ad omnem Mathematicæ | scientiæ partem, tùm ad quamlibet | Geometriæ tractationem, fa-|cilis comparatur | aditus. || [DEVICE.] | COLONIÆ, | Apud Maternum Cholinum. | M.D.LXXXVII. | Cum Gratia *&* Priuilegio Cæs. Maiest.

Octavo. A–O⁸ P⁴. 116 leaves. Pagination, after 14 unnumbered leaves, 2–203. Catchwords. 3ᵃ: 31 lines, 118 (with headline 122) × 64 mm. Types: text roman, preface italic, some greek. Ornamental capital on 2ᵃ. Diagrams with floral ornaments in text.

Riccardi.

B.M.

III

TRANSLATIONS INTO CURRENT EUROPEAN LANGUAGES, AND INTO ARABIC

No. 34

del sesto) quello che peruiene dalla.h.in la.g.d.e equale a quello che uien fatto dalla.a.b.in la.d,e.Et (per el correlario della precedente) e manifesto che la proportione de tutte le superficie del dodecedro(del quale el lato e la.a.b.(tolte insieme,a tutte le superficie del ycocedro(del quale el lato e la.a.c.)tolte insieme,e si come di quello che uien fatto dalla.a.b.in la.d.e.a quello che uien fatto dalla a.c.in la.g.d.Adonque(per la prima parte della settima del quinto,&undecima del medesimo)la proportione di quello che peruiene dalla.h.in la.g.d . a quello che peruiene dalla.a.c.in la.g.d.e si come de tutte le superficie di quel dodecedron a tutte quelle di questo ycocedro:Ma di quello che peruiene dalla.h.in la g.d.a quello che peruiene dalla.a.c.in la.g.d.(per la prima del sesto) e si come della.h.alla.a.c.Adonque(per la.11.del.5.la proportione di tutte le superficie di quel dodecedro a tutte quelle di questo ycocedron e si come della . h . alla.a.c. che e il proposito , Questo medesmo poteremo prouare altramente: se auanti quello poneremo un antecedente necessario el qual e questo.

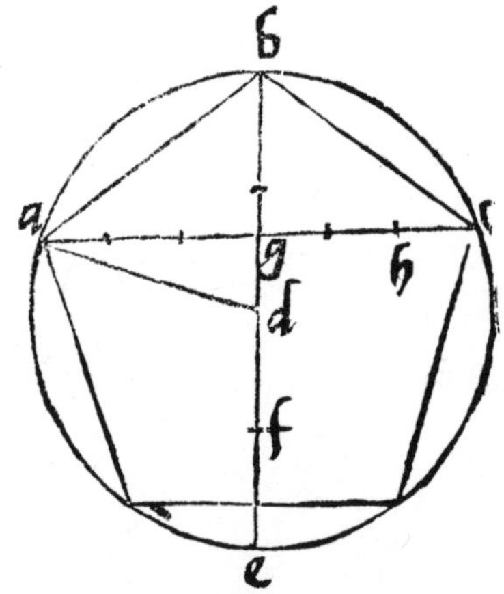

Se in qualunque cerchio sara inscritto un penthagono equilatero lo rettangolo che e cõtenuto sotto il dodrante del diametro di quel cerchio & sotto dextante di quella linea che tende sotto al angolo di quel penthagono de necessita el bisogna esser equale al medesimo penthagono.

Li nostri maggiori con lo intelletto & con la ragione diuiderno cadauno integro in dodeci parti equali & tutte q̃lle p arti insieme(cioe quel tutto)lo chiamorno. Asse & le undice di quelle parti gli dissono deunce . Et le diece, dextante: le noue dodrante,& le otto bisse,& le sette,sseptunce ouer septante ouer quincũce & le sei:semis,& le cinque quincunce & le quatro triente:& le tre, quadrante: & le due sextante,& la una,adimandorno oncia,& quelle piu uolte sono sta trouate in li antiqui libri designate per lordine de tai figure.

As Deunce Dextante Dodrante Bisse Septunce

Semis. Quincunce. Triente. Quadrante Sextante. Vncia.

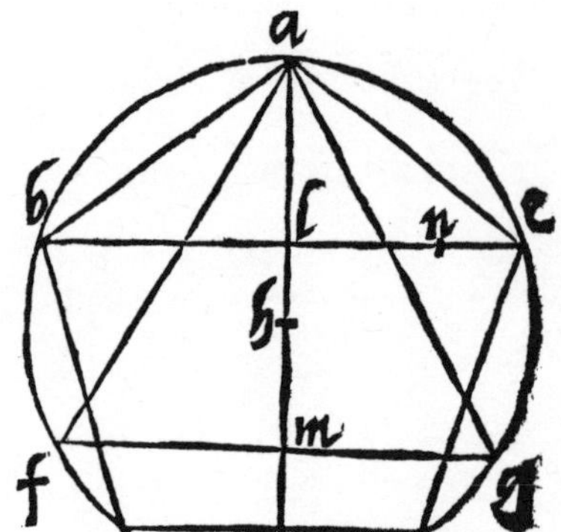

Anchora la onza la qual hauemo detto douer esser la.12.parte del As.la diuiderno in altre.12.frattioni,ma p unaltra uia,pche la mita della onza gli dissono, Semioncia. La terza parte duella.la quarta sicilico.la sesta:sextula.la ottaua dragma la duodecima emissela.la.18,tremisse.la.24.scrupulo.la. 48 . obolo . la 72.Bissiliqua.la.96.cerates.la ultima chi e la.144.pte di essa oncia chiamorno siliqua, Et a queste.12.frattioni della oncia li posteriori,gli hãno agiõto el calco & lo carco e la.192.pte della oncia:del qual agiongimẽto ne fu causa accio che el diatesseron & el diapente delle simphonie di toni & semitoni distinti p interuali di queste frattioni.la denominatione ascendesse ouer se estendesse per fina al minimo istremo:& tutte q̃lle frattioni li anotauano secondo lordine de ta figure.

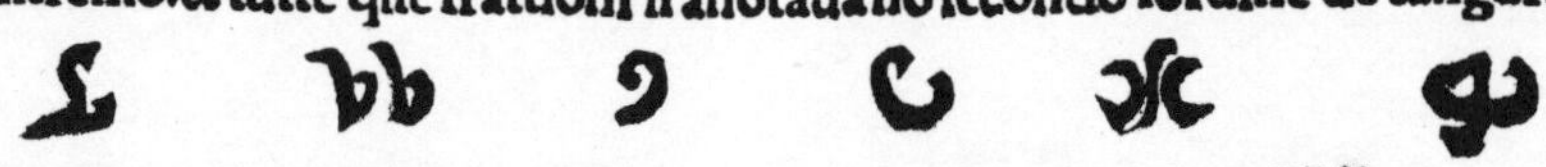

Semiuncia. Duella. Sicilico. Dragma. Emissella.

Tremisse Scrupulo. Obolo. Bissiliqua. Cerates. Siliqua. Calco.

34.

1543, February. Venice. Venturino Roffinelli.

1ª. EVCLIDE MEGARENSE | PHILOSOPHO: | SOLO INTRODVTTORE | DELLE SCIENTIE MATHEMATICE: | DILIGENTEMENTE REASSETTATO, ET ALLA | INTEGRITA RIDOTTO PER IL DEGNO | Professore di tal Scientie Nicolo Tartalea, | BRISCIANO, | Secondo le due Tradottioni: | E PER COMMVNE COMMODO | & vtilita di latino in volgar | tradotto. | CON VNA AMPLA ESPOSITIONE | DELLO ISTESSO TRADOTTORE | DI NOVO AGGIONTA. | Talmente chiara, che ogni mediocre ingegno, senza la notitia, ouer suffra=|gio di alcun' altra scientia con facilita, sera capace a poterlo intendere. || [DEVICE with letters G.T.] 241ª. Stampato in Vinegia per Venturino Roffinelli ad instantia e requisi|tione de Guilielmo de Monferra, & de Pietro di Facolo da | Vinegia libraro, & de Nicolo Tartalea Brescia=|no Tradottore: Nel Mese di Febraro. | Anno di nostra salute || M.D.XLIII. || Con gratia & priuilegio . . . come nel priuilegio si contiene. || [DEVICE with letters .P.Z.F.] 241ᵇ. [Register.] 242ᵇ. [DEVICE with letters .P.Z.F.]

Folio. A^{10} B–Z & ꝯ ℞ $AA–DD^{8}$. 242 leaves, 2–241 numbered II–CCXXXIX with errors and omissions. Catchwords. 24ª: 50 lines (mixed types and spaces), 231 (with headline 240) × 125 (with diagrams 175) mm. Type: roman. Ornamental capitals. Marginal diagrams.

The sheet signed Av containing a list of errata was evidently inserted as an afterthought, as the pagination is interrupted thereby.

The first translation in any modern language, reprinted in 1565, 1569, 1585, 1586. The two translations used by Tartalea are those of Campanus and Zamberti.

Reprints of 1544 and 1545 are recorded by Argelati, *Bib. de volgarizzatori*, ii, p. 41, and also by Graesse.

Riccardi.

B.M. Bodl. C.T–S.

35.

1545. Rome. Antonio Blado.

I QVINDICI | LIBRI DEGLI ELEMEN|TI DI EVCLIDE, DI GRE|CO TRADOTTI IN | LINGVA THO=|SCANA. | [woodcut portrait of Euclid] | IN ROMA. MDXXXXV. || Con gratia & priuilegio del S. N. S. Paulo Terzo, | & della Serenissima republica Venetiana | per cinque anni. 55ᵇ. Stampata in Roma per Antonio Blado Asolano. | MDXLV.

Octavo. $A–G^{8}$. 56 leaves, the last blank. Pagination, after 2 unnumbered leaves, 5–107. 6ª: 26 lines, 106 (with book number 110) × 68 mm. Types: text and prefatory matter italic, title and book numbers roman capitals. Ornamental capital on 2ª.

A companion volume to the Greek text issued by the same printers in the same year. Additional matter on 55ª—Le tredici rette Irrationali, che Euclide | tratta nel decimo libro—shows it to be probably the later of the two.

Riccardi.

B.M. Bodl. C.T–S. G

36.

1562. Basle. Jacob Kündig.

(red) Die Sechs Erste Bůcher | Euclidis, | Vom anfang oder grund | der Geometrj. || (black) In welchen der rechte grund, nitt allain der Geometrj | (versteh alles kunstlichen, gwisen, vnd vortailigen ge⸗|brauchs des Zirckels, Linials oder Richtscheittes vnd | andrer werckzeüge, so zů allerlaj abmessen dienstlich) | sonder auch der fürnemsten stuck vnd vortail | der Rechenkhunst, furgeschriben vnd | dargethon ist. || (red) Auß Griechischer sprach in die Teütsch gebracht, aigent⸗|tlich erklårt, Auch mit verstentlichen Exempeln, gründ⸗|lichen Figurn, vnd allerlaj den nutz für augen stellen⸗|den Anhången geziert, Dermassen vormals | in Teütscher sprach nie gesehen | worden. || (black) Alles zů lieb vnd gebrauch den kunstliebenden Teütschen, so sych der Geo⸗|metrj vnd Rechenkunst anmassen, mit vilfåltiger mühe vnd arbait | zum trewlichsten erarnet, vnd in Truckh ge⸗|geben, Durch || (red) Wilhelm Holtzman, genant Xylander, | von Augspurg. || (black) Getruckht zů Basel. 105ª. Vollendet durch Iacob Kündig, zů Basel, in | Ioanns Sporini kosten, im jar 1562. | auff den dreyßigsten tag des | Winmonats.

Folio. a^4; b^4; a^4; A–O^6 P^4 Q^6. 106 leaves, 8, 106 blank. Pagination, after 12 preliminary leaves, 1–185, followed by blank page and leaf. Catchwords. 3^a: 43 lines, 237 (with headline 247) × 127 (with marginalia 140) mm. Types: gothic; some greek and italic (marginalia). Ornamental capitals and tailpieces. Diagrams in text.

Holtzman's preface is dated 'Heydelberg, Anno 1562, den ersten Octobris'.

Riccardi.

C.T–S. U.L.C.

See Pl. 9.

37.

1564. Paris. H. de Marnef & G. Cavellat.

LES SIX | PREMIERS LIVRES | DES ELEMENTS D'EV-|CLIDE, TRADVICTS ET | COMMENTEZ PAR PIERRE | Forcadel de Bezies, Lecteur ordi-|naire du Roy és Mathema-|tiques en l'vniuersité | de Paris. | *** || [DEVICE with type-set motto.] || A PARIS, || Chez Hierosme de Marnef, & Guillaume Cauellat, | au mont S. Hilaire, à l'enseigne du Pelican. || 1564. 192^b. [DEVICE.]

Quarto. $*^2$; A–Z AA–ZZ Aa^4 Bb^2. 192 leaves, 3–192 numbered 1–190. Catchwords at ends of quires. 13^b: 35 lines (mixed type), 167 (with headline 171) × 99 mm. Types: enunciations italic, demonstrations roman. Ornamental capitals and headpieces. Diagrams with fleurons in text.

Many copies want the last cipher of the date. In the Libri copy, now in the British Museum, 9 has been added in pen and ink. Hence the edition of 1569 quoted by Brunet.

Riccardi.

B.M. C. T–S.

See Pl. 10.

38.

1565. Paris. Charles Perier.

LES | SEPTIEME, HVICTIEME | ET NEVFIEME, LIVRES | DES ELEMENS D'EVCLIDE, | COMPRENANS TOVTE LA SCIEN-|CE des nombres, traduits & com-|mentez par Pierre Forcadel de | Beziés, lecteur ordinaire | du Roy és Mathema-|tiques, en l'vni-|uersité de | Paris. || [DEVICE.] || A PARIS. | Chez Charles Perier, demourant en la rue | S. Iean de Beauuais, au Bellerophon. | 1565. | AVEC PRIVILEGE DV ROY.

LIBRO QVARTO DI EVCLIDE.

Diffinitione prima.

1/1 VNA figura rettilinea uiene detta esser descritta in un'altra figura rettilinea, quando ciascun'angolo della figura inscritta tocca ciascun lato de quella in laquale è descritta.

Sia il triangolo, a,b,c, descritto di dentro del triangolo,d, e, f, talmente che ciascun angolo del triangolo,a,b,c, tocca ciascun lato del triangolo, d,e,f, (in li tre ponti,a,b, c,) hor dico che'l triangolo,a,b,c, uien detto esser inscritto in lo triangolo,d,e,f,similmente sel fusse il quadrato,a,b,c,d,descritto di dentro dil quadrato,e,f,g, h,talmente che ciascun angolo del quadrato,a,b,c,d, tocchi ciascũ lato del quadrato,e,f,g,h,(nelli quattro ponti,a,b,c,d,)dico che il quadrato,a,b,c,d, uien detto esser inscritto di dẽtro del quadrato,e,f,g,h, et cosi si deue intendere de ogni altra sorte de figura contenuta de linee rette.

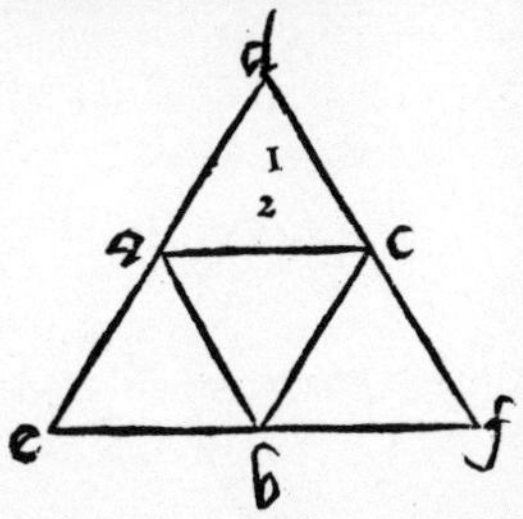

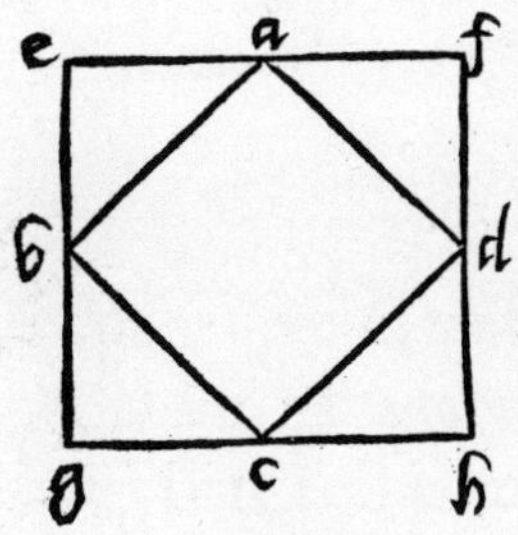

Diffinitione. 2.

2/2 Similmente una figura uien detta esser descritta cerca a un'altra figura, quando ciascuno lato della circonscritta tocca ciascun angolo de quella cerca laquale è descritta.

Sia come è il triangolo, d, e, f, (della precedente) che ciascun lato di quella tocca ciascũ angolo del triãgolo,a,b,c.per laqualcosa il triangolo.d.e.f.uien detto esser descritto attorno al triangolo,a,b,c, & similmẽte il quadrato,e,f,g,h,uien detto esser descritto cerca al quadrato,a,b,c.d,perche ciascuno lato di quello tocca ciascuno angolo del detto quadrato,a,b,c,d.

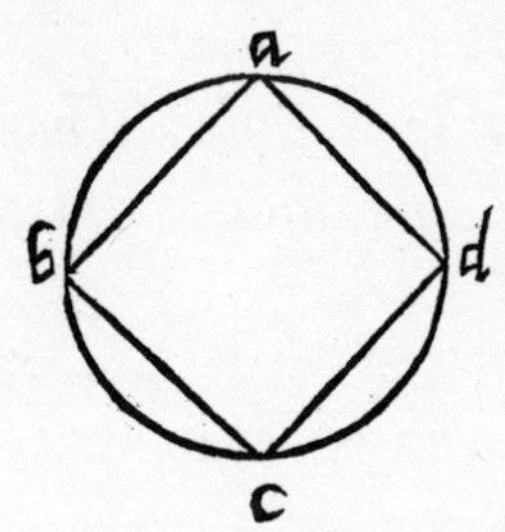

Diffinitione. 3.

Vna figura rettilinea uiẽ detta esser descrit
o/3 ta in uno cerchio, quando ciascaduno angolo della inscritta tocca la circon ferentia dello cerchio.

Si come

The firſt Probleme. The firſt Propoſition.

Vpon a right line geuen not beyng infinite, to deſcribe an equilater triangle, or a triangle of three equall ſides.

SVppoſe that the right line geuen be A B. It is required vpon the line A B, to deſcribe an equilater triangle, namely, a triangle of three equall ſides. Now therfore making the centre the point A Conſtruction. *and the ſpace A B, deſcribe (by the third peticion) a circle B C D and agayne (by the ſame) makyng the centre the point B, and the ſpace B A, deſcribe an other circle A C E. And (by the firſt peticion) from the point C, wherin the circles cut the one the other, draw one right line to the point A, and an other right line to the point B. And foraſmuche as the point A is the centre of the circle C B D, therfore (by the 15. definition) the line A C is equall to the line A B: Agayne foraſmuch as the point B is the centre of the circle C A E, therfore (by the ſame definition) the line B C is equall to the line B A. And it is proued, that the line A C is equall to the line A B: wherfore either of theſe lines C A and C B, is equall to the line A B: but thinges which are equall to one and the ſame thing, are alſo equall the one to the other (by the firſt common ſentence) wherfore the line C A, alſo is equall to the line C B. VVherfore theſe three right lines C A, A B, and B C are equal the one to the other. VVherfore the triãgle A B C is equilater. VVherfore vppon the line A B, is deſcribed an equilater triangle A B C. VVherfore vppon a line geuen not being infinite, there is deſcribed an equilater triangle, VVhich is the thing, which was required to be done.*

Demonſtration

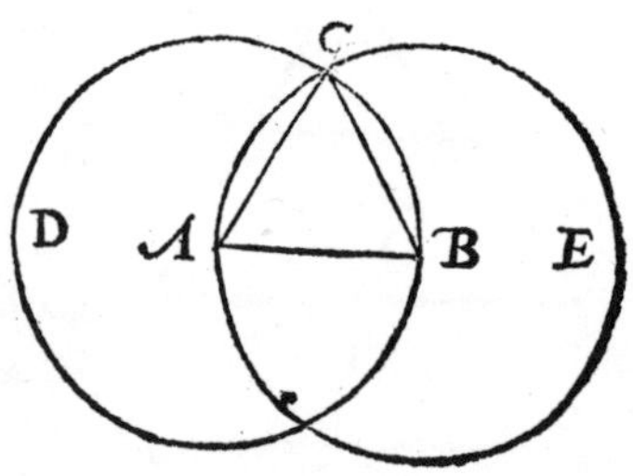

A triangle or any other rectilined figure is then ſaid to be ſet or deſcribed vpon a line, when the line is one of the ſides of the figure.

This firſt propoſition is a *Probleme*, becauſe it requireth acte or doyng, namely, to deſcribe a triangle. And this is to be noted, that euery *Propoſition*, whether it be a *Probleme*, or a *Theoreme*, commonly containeth in it *a thing geuen, and a thing required to be ſearched out:* although it be not alwayes ſo. And the *thing geuen*, is euer ſet before the *thing required*. In ſome propoſitions there are more things geuẽ then one, and mo thinges required then one. In ſome there is nothing geuen at all.

Thing geuen. *Thing required*

Moreouer euery *Probleme & Theoreme*, beyng perfect and abſolute, ought to haue all theſe partes, namely. Firſt the *Propoſition*, to be proued. Then the *expoſition*

Propoſition. *Expoſition.*

Quarto. A–O⁴ P⁶. 62 leaves, the last blank. Pagination, after unnumbered title-leaf, 3–122. 6ᵃ: 36 lines (mixed type), 162 (with headline 168) × 100 mm. Types: enunciations italic, demonstrations roman. Ornamental capitals and headpieces.

Riccardi.

B.M. Bodl.

39.

1565/6. Venice. Curtio Troiano Navo.

EVCLIDE | MEGARENSE | PHILOSOPHO, | SOLO INTRODVTTORE | DELLE SCIENTIE | MATHEMATICE. || DILIGENTEMENTE RASSETTATO ET ALLA | integrità ridotto, per il degno professore di tal Scientie | Nicolo Tartalea Brisciano. || SECONDO LE DVE TRADOTTIONI. || CON VNA AMPLA ESPOSITIONE | dello istesso tradottore di nuono aggiunta. || TALMENTE CHIARA, CHE OGNI MEDIOCRE | ingegno, senza la notitia, ouer suffragio di alcun' altra scientia | con facilità serà capace a poterlo intendere. || [DEVICE.] || IN VENETIA, Appresso Curtio Troiano. 1565. REGISTRO. || . . . || [DEVICE.] || IN VENETIA, | APPRESSO CVRTIO TROIANO. | M.D.LXVI.

Quarto. A–Z Aa–Qq⁸. 312 leaves, 2–120 so numbered, 121–311 numbered 125–315. 4ᵃ: 42 lines, 172 (with foliation 176) × 110 mm. Types: enunciations roman, demonstrations italic. Ornamental capitals. Diagrams in text.

Reprinted in 1585, 1586.

Riccardi.

B.M. Bodl. C.T–S.

40.

1569. Venice. Giovanni Bariletto.

EVCLIDE MEGARENSE | ACVTISSIMO PHILOSOPHO, | SOLO INTRODVTTORE DELLE | SCIENTIE MATHEMATICE. | DILIGENTEMENTE RASSETTATO, ET ALLA | integrità ridotto, per il degno professore di tal Scientie | Nicolò Tartalea Brisciano. || SECONDO LE DVE TRADOTTIONI. || CON VNA AMPLA ESPOSITIONE | dello istesso tradottore di nuouo aggiunta. || TALMENTE CHIARA, CHE OGNI MEDIOCRE | ingegno, senza la notitia, ouer suffragio di alcun' altra scientia | con facilità serà capace a poterlo intendere || [DEVICE.] | IN VENETIA, Appresso Giouanni Bariletto. 1569. REGISTRO. || . . . | Tutti sono quaderni. || [DEVICE.] || IN VENETIA, | Appresso Giouanni Bariletto. | M.D.LXIX.

Quarto. A–Z Aa–Qq⁸. 312 leaves, 2–311 numbered 2–120, 125–315. 6ᵃ: 42 lines, 172 (with foliation 176) × 110 mm. Types: roman, italic, prefatory matter roman. Ornamental capitals. Diagrams in text.

Riccardi.

C.T–S.

41.

1570. London. John Day.

THE ELEMENTS | OF GEOMETRIE | of the most aunci-|ent Philosopher | EVCLIDE | of Megara. || Faithfully (now first) tran=|slated into the Englishe toung, by | H. Billingsley, Citizen of London. | Whereunto are annexed certaine | Scholies, Annotations, and Inuenti-|ons, of the best Mathematici-|ens, both of time past, and | in this our age. ||

With a very fruitfull Præface made by M. I. Dee, | specifying the chiefe Mathematicall Sciēces, what | they are, and wherunto commodious: where, also, are | disclosed certaine new Secrets Mathematicall | and Mechanicall, vntill these our daies, greatly missed. || Imprinted at London by Iohn Daye. 491^{b}. [DEVICE.] | AT LONDON | Printed by Iohn Daye, dwelling | ouer Aldersgate beneath Saint Martins. || ℂ These Bookes are to be solde at | his shop vnder the gate. || 1570.

Folio. ☞ * a–d A^4; B–X^4 Aa–Yy AA–YY AAa–FFf^6 GGg–YYy AAA–EEE^4. 492 leaves, the last blank, 30–231 numbered 2–203, 232–491 numbered 205–464. 5^a: 49 lines, 237 (with headline 246) × 140 (with marginalia 162) mm. Types: italic, roman. Ornamental capitals. Woodcut border with figures, signed IB, enclosing title. Diagrams in text. Device, McKerrow 145.

Contents: 1^a, title within border; 1^b, blank; 2^a, translator's preface; 3^b, blank; 4^a, Mathematicall Præface of John Dee, followed by folding 'Groundplat'; 29^a, Euclid's Elements; 444^b, Sixtenth booke of the Elementes of Geometrie added by Flussas; 485^b, Flussas of mixt and composed regular solides; 490^h, Faultes escaped; 491^b, Device and colophon; 492, blank.

Riccardi.

B.M. Bodl. C.T–S.

See Pl. 11.

42.

1575. Urbino. Domenico Frisolino.

DE GLI | ELEMENTI | D' EVCLIDE | LIBRI QVINDICI. || CON GLI SCHOLII ANTICHI. || TRADOTTI PRIMA | IN LINGVA LATINA DA M. | FEDERICO COMMANDINO DA VRBINO, | & con Commentarij illustrati, || ET HORA D'ORDINE DELL' ISTESSO | TRASPORTATI NELLA NOSTRA VVLGA-|re, & da lui riueduti. || CON PRIVILEGIO. || JN VRBINO, APPRESSO | DOMENICO FRISOLINO. | M.D.LXXV. | Con licentia de' Superiori. 286^a. JN VRBINO IN CASA DI FEDERICO | COMMANDINO, CON LICENTIA DE' SVPERIORI. | M.D LXXV.

Folio. [$*^2$] $**^4$ $***^2$; A–Z AA–ZZ AAa–ZZz^4 $AAAa^2$. 286 leaves, 9–286 numbered 1–278. Double rules round type-page enclosing headlines and marginalia. 5^a: 53 lines, 222 × 127 (with rules 238 × 149) mm. Types: enunciations roman, demonstrations italic. Ornamental capitals. Diagrams in text.

Edited by Valerio Spacciuoli.

Riccardi.

B.M. C.T–S.

43.

1576. Seville. Alonso de la Barrera.

LOS SEIS LIBROS | PRIMEROS DELA GEOMETRIA | DE EVCLIDES. | Traduzidos en lēgua Española por Rodrigo çamorano Astrolo|go y Mathematico, y Cathedratico de Cosmographia por | su Magestad en la casa de la Contrataciõ de Seuilla | Dirigidos al jllustre señor Luciano de Negrõ, | Canonigo dela sancta yglesia de Seuilla | || [Coat of Arms] || Con licencia del Consejo Real. | En Seuilla en casa de Alonso de la Barrera. | ❧ 1576. ❧ || Esta tassado en

DE GLI ELEMENTI DI EVCLIDE
LIBRO SESTO
CON LI SCHOLII ANTICHI, ET COMMENTARII
Di Federico Commandino da Vrbino.

DIFFINITIONI.
I.

LE figure rettilinee simili sono quelle, che hanno ciascun'angolo vguale, & d'intorno à gli vgual angoli i lati proportionali.

II.

Le figure reciproche sono, quando nell'vna & nell'altra siano le antecedenti & conseguenti proportioni.

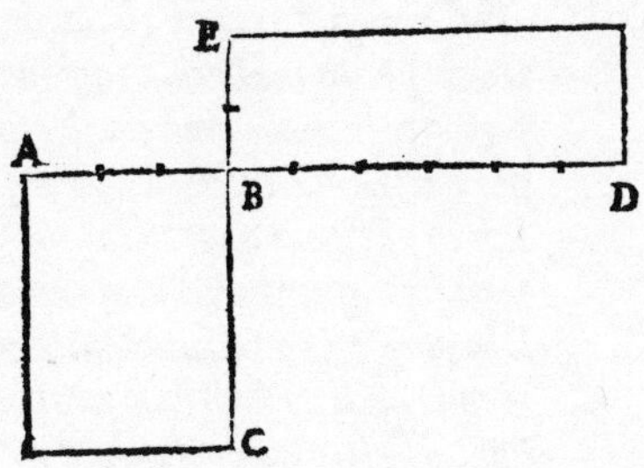

IL COMMANDINO.

Per le antecedenti & conseguenti proportioni, intendi i termini antecedenti & conseguenti delle proportioni, come se siano due rettangoli ABC DBE, & sia come la AB alla BD, così la EB alla BC, si diranno queste figure essere reciproche, ò vero rispondenti fra loro contrariamente. percioche nell'vna è

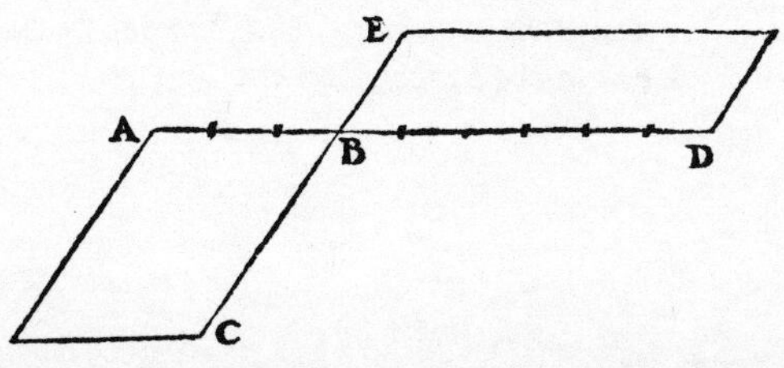

il

la superficie,e,serà uno.64.esimo & uno 256.esimo.(cioe in summa cinque.256.esimi)sia adonque.f.lo lato tetragonico della b.& la.g.sia el lato tetragonico del secondo lato della detta superficie.b &(per lo premesso antecedente)serà che dal.f.in g. sia fatto. a. linea cioe R. 5. un'altra uolta sia la:h.lo lato tetragonico del secõdo lato della superficie.c.& sia anchora.k. el lato tetragonico de,h. & per lo predetto antecedẽte serà che dal.b.in,b.sia fatto.a.& dal.f. in k.sia fatto el lato tetragonico de a.qual sia.l. sia ancora m.lo lato tetragonico del secondo lato della superficie. d.& quando che.n,sia el lato tetragonico de m.& p.el lato tetragonico de n.&(per lo predetto antecedente)serà che dal. c.in m,sia fatto a.& dal.b.in.n:sia fatto l.& dal.f.in p.sia fatto el lato tetragonico de l.(qual sia.q.)ma piu sia.r. el lato tetragonico del secondo lato della superficie,e,ancora sia s.lo lato tetragonico de.r.& t. lo lato tetragonico de.s.& .u.sia lo lato tetragonico de. t.& seguita (per lo detto antecedente)che dal.d.in.r.sia fatto.a.et dal c.in.s,sia fatto l.& dal b.in t.sia fatto.q.& etiã da l, f. in u. sia fato el lato tetragonico de q. (qual sia.x.) & cosi in infinito. Dico adonque queste linee. a.l. q.x.(dellequal la.a.è come radical principio) esser irrationale la:a. solamente in lõghezza, tutte le altre in longhezza & in potẽtia, & dico che niuna di quelle conuien con alcun'altra in diffinitione, ouer in ordine.perche conciosia che dal: f.in.g. & k.uengono fatti.a.& .l.serà dal.a.al.l:sicome dal.g.al.k.& perche(come è manifesto dalli detti

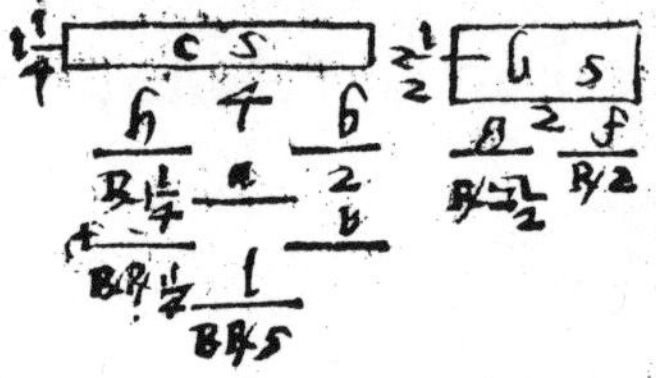

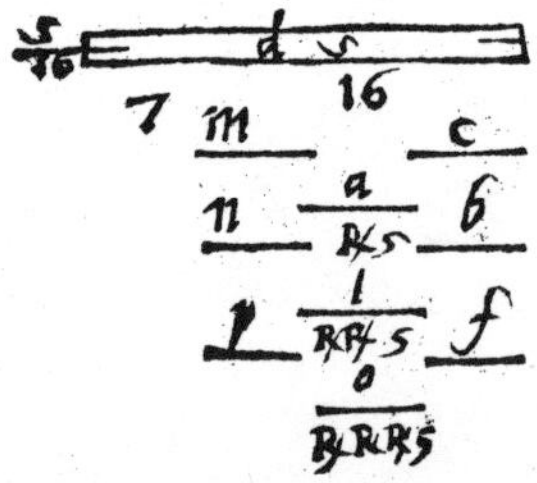

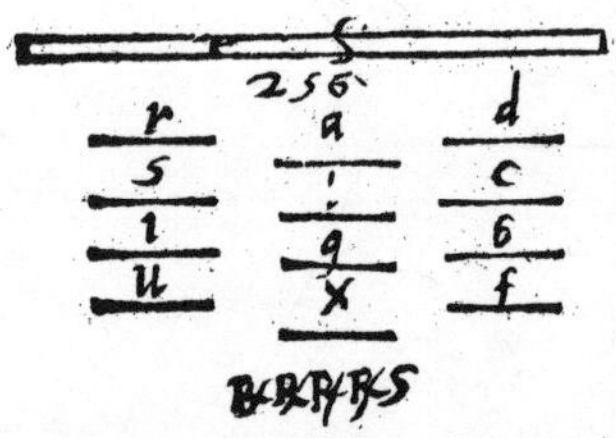

$\frac{5}{256}$ presuppositi)g.& K.sono incommensurabili in longhezza & in potentia seguita etiam che.a.& .l.siano incommensurabili in lõghezza & in potẽtia & per la medesima ragione etiam.d.q. perche dal.a.al; q. e si come dal.g. al p.& per la medesima causa etiam. a.& .x.conciosia che siano sicome.g.& .u.& per questa uia anchora è necessario che l.& q.siano similmente incommensurabili si in potentia quanto in longhezza,perche conciosia che dal.f.in K.& p. siano fatti.l.& .q.serà del.l.al.q.come del.K.al.p.ma. K.e. p. non sono commensurabili in longhezza ne inpotentia.perche essendo commensurabili.h.& .n.seriano cõmensurabili & non sono,anchora,l,&,x,è necessario esser incommensurabili in l'uno e l'altro modo perche dal,l,al.x.e sicome dal. k.al.u. imperoche dal.f. in.K. & u.sono fatti,l,& x.& k.&,u,sono incommensurabile in l'uno & l'altro modo,perche ponendo che fusseno per l'aduersario seguiria,t,&,b,asser commensurabile che è inconueniente.

Quarto. A^4 B C–P^8 Q^4 R^2. 122 leaves, 4–121 so numbered. 5^a: 25 lines, 151 × 94 mm. Types: roman. Ornamental capitals. Diagrams in text.

Riccardi.

B.M. C.T–S.

44.

1585. Venice. Heirs of Troiano Navo.

EVCLIDE MEGARENSE | ACVTISSIMO PHILOSOPHO | SOLO INTRODVTTORE DELLE | SCIENTIE MATHEMATICE. || DILIGENTEMENTE RASSETTATO, | Et alla integrità ridotto, per il degno professore di tal | Scientie Nicolò Tartalea Brisciano. | SECONDO LE DVE TRADOTTIONI. | CON VNA AMPLA ESPOSITIONE DELLO ISTESSO | tradottore di nuouo aggionta, talmente chiara, che ogni mediocre | ingegno, senza la notitia, ouer suffragio di alcun' altra scientia | con facilità, serà capace a poterlo intendere. | Di nuouo con ogni diligenza ben corretto, e ristampato. || [DEVICE.] || IN VENETIA, | Appresso gli Heredi di Troian Nauo, alla libraria dal Lione. | M D LXXXV.

Quarto. A–Z Aa–Qq^8. 312 leaves, 3–120 so numbered, 121–311 numbered 125–315. Catchwords. 4^a: 42 lines, 174 (with foliation 177) × 109 mm. Types: Lettione of Tartalea roman, text italic. Ornamental capitals. Diagrams in text.

A reprint of the 1565–6 edition with the error in pagination repeated and with a different dedicatory letter.

Riccardi.

B.M.

45.

1586. Venice. Heirs of Troiano Navo.

EVCLIDE MEGARENSE | ACVTISSIMO PHILOSOPHO | SOLO INTRODVTTORE DELLE | SCIENTIE MATHEMATICE. | DILIGENTEMENTE RASSETTATO, | Et alla integrità ridotto, per il degno professore di tal | Scientie Nicolò Tartalea Brisciano. | SECONDO LE DVE TRADOTTIONI. | CON VNA AMPLA ESPOSITIONE DELLO ISTESSO | tradottore di nuouo aggionta, talmente chiara, che ogni mediocre | ingegno, senza la notitia, ouer suffragio di alcun' altra scientia | con facilità serà capace a poterlo intendere. | Di nuouo con ogni diligenza ben corretto, e ristampato. || [DEVICE.] || IN VENETIA, | Appresso gli Heredi di Troian Nauo, alla libraria dal Lione. | M D LXXXVI. 312^b. REGISTRO. | . . . Quaderni. || [DEVICE.]

Quarto. A–Z Aa–Qq^8. 312 leaves, 3–120, so numbered, 121–311 numbered 125–315. Catchwords. 26^a: 42 lines, 172 (with headline 176) × 110 mm. Types: italic, roman, prefatory matter roman. Ornamental capitals. Diagrams in text.

Paris, Bib. Nat.

46.

1594. Rome. Typographia Medicea.

كتاب تحرير اصول لاوقليدس
من تاليف خوجه
نصير الدين الطوسي

EVCLIDIS | ELEMENTORVM | GEOMETRICORVM | Libri Tredecim. || EX TRADITIONE DOCTISSIMI | NASIRIDINI TVSINI || Nunc primum Arabicè impressi. || ROMÆ || In Typographia Medicea. | M.D.XCIV. | Cum licentia Superiorum.

Folio. 228 leaves, the last blank, with pagination 2–453, 3–6, 9–24 being omitted and 8 replacing 7. Catchwords. 5[a]: 35 lines, 231 (with headline 248) × 119 mm. Types: arabic, except part of title in roman. Book-titles printed from blocks. Type-page enclosed with double rules. Diagrams in text.

Riccardi.

B.M. Bodl. C.T–S.

46 a.

[1594 ?] [Rome. Typographia Medicea.]

Another issue:—

كتاب تحرير اصول لاوقليدس
من تاليف خوجه
نصير الدين الطوسي

In this issue the first quire (4 leaves) has been reset, omitting the roman portion of the imprint.

Riccardi.

B.M.

See Pl. 12.

IV

FRAGMENTARY EDITIONS IN VARIOUS LANGUAGES

The translation of Euclid by Boethius is included in the following editions of the Opera:

1492, 18 August. Venice. Joannes et Gregorius de Gregoriis. Hain *3351. Proctor 4517. B.M.C. V. 341.

1499, 8 July. Venice. J. & G. de Gregoriis. Hain *3352. Proctor 4559. B.M.C. V. 351.

1546, March. Basle. Heinrich Petri.

1570, March. H. Petri.

Boethius' translation also occurs in the following editions of the *Sphaera Mundi* of J. de Sacro Bosco:

1500, 1 September. Paris. Wolfgang Hopyl. fol.

1507, 10 November. Paris. Henri Estienne. fol.

1516, 10 May. Paris. Henri Estienne. fol.

[1520?] Avignon. Joannes de Channey. 4°.

1521, 26 April. Paris. Simon de Colines. fol.

1527, 21 August. Paris. S. de Colines. fol.

[1531. Paris. S. de Colines. fol. Murhard, ii, p. 21.]

1534. Paris. S. de Colines. fol.

1538, 13 March. Paris. fol.

I.

1506. Lib. 1–4 Elementorum cum Campani commentario. Frankfort.

1^a. Euclides. 46^b (last par.). Et tantuʒ de quatuor libris elementoꝝ Euclidis | cum familiari Campani in eosdeʒ comentario. qui | cuʒ in omnibꝰ vniuersalibꝰ studijs ordinaria institu|cione preeleguntur sic eciaʒ in hac nostra achade⸗|mia franckfordiana. ab illustrissimis Ioachim prin|cipe electore ⁊ Alberto . . . nuper electa. per Mgrʒ Ambrosiuʒ lach|er de Merspurg mathematicuʒ | ibidʒqʒ Collegiatuʒ | accurata diligencia lecti castigati propriisqʒ impen|sis elaborati sunt atqʒ impressi ut athenaꝝ more | studium philosophie a mathematico splendore mi|rifice inceptuʒ efflorescat Anno salutis. vi.

Quarto. A–H$^{6.4}$ I^6. 46 leaves. 13^a: 43 lines, 147 (with headline 157) × 70 (with diagrams 126) mm. Types: gothic. Lombard capitals on 1^a, capital spaces elsewhere. Marginal diagrams.

Riccardi.

B.M.

II.

1528. Elementale Geometricum a I. Voegelino ex Euclidis Geometria. Vienna. J. Singrenius.

Elementare geometricum . . . In ædibus Io. Singrenii.

No copy of this work has come to the notice of the compiler of this bibliography, but there seems no reason to doubt its existence.

Quarto.

Murhard, ii, p. 21, Fabricius, Riccardi.

III.

1529. Elementale Geometricum ex Euclidis geometria a I. Voegelin decerptum. Strasburg. Christian Egenolff.

❧ ELEMEN|TALE GEOMETRICVM, EX EVCLI|dis Geometria, à Ioanne Vœgelin, Haylpronnensis | ad omnium Mathematices studiosorum | utilitatem decerptum. || GERARDVS NOVIOMAGVS | Lectori. | Quid punctum, quid linea, circulus, angulus, orbis | Designet, paucis hæc elementa docent. | Sydereos cursus hinc disces, ut uenerere | Patrem, qui terras condidit. astra, polos. || [DEVICE, with legend: IN SVDORE VVL|TVS TVI VES⸗||CERIS PA⸗|NE TVO. Gen. 3.] | ARGENTORATI, apud Christianum | Egenolphum. An. M. D. XXIX.

Octavo. A–C^8 D^6. 30 leaves. 5^a: 28 lines, 119 x 65 (with diagrams 82) mm. Catchwords. Types: italic. Marginal diagrams.

B.M.

IV.

1529. Giambattista Politi. Expositio super definitiones. Siena. Simone Nicolō de' Nardi.

Expositio Io. Bapt. Politi super definitiones et propositiones quae supponuntur ab Euclide in quinto Elementorum eius.

Octavo.

Riccardi. *Bib. mat.* ii, col. 298.

V.

1530. Thomas Bradwardinus. Geometria speculatiua (including the Elementale Geometricum of Joannes Voegelin). Paris. Regnault Chaudière.

❧ GEOMETRIA SPE|CVLATIVA THOME BRAVARDINI | recoligens omnes conclusiones geome⸗|tricas studentibus artium & | philosophie aristotelis val⸗|de necessarias simul | cum quodam tra|ctatu de qua⸗|dratura cir|culi no⸗|uiter | edi⸗|to ∴ || ℂ Elementale geometricum ex euclidis geo|metria a Ioanne Voegelin Haylpronnensi | ad omnium mathematices studiosorum vti⸗|litatem decerptum. || Parisijs apud Reginaldum Chauldiere sub | homine Siluestri in vico diui Iacobi.

Folio. A B^6 C D^4; E^6 F^4. 30 leaves, 2–30 so numbered. 3^a: 53 lines, 211 (with foliation 215) × 113 (with diagrams 167) mm. Types: roman. Ornamental capitals. Marginal diagrams.

1^a, title; 1^b, blank; 2^a, Breue compendium artis geometrie ex libris Euclidis Bœcij & cāpani cōpilatū; 20^a, tractatus de quadratura circuli; 21^a, Elementale Geometricum; 30^a, colophon; 30^b, blank.

B.M.

VI.

1534. Iohannes Wolphius. Rudimenta Arithmetices (including the Elementale Geometricum). Frankfort. Christian Egenolff.

☙ RVDI|MENTA ARITHMETICES | Authore Iohanne Vuolphio | Hersbrugiense. || ELEMEN|TALE GEOMETRICVM, EX | Euclidis Geometria, à Ioanne Vœgelin, | Haylpronnensi, ad omnium Mathe|matices studiosorum utili⸗|tatem decerptum. || FRANC. Christianus Ege⸗|nolphus excudebat.

Octavo. A–G^8. 56 leaves. 10^a: 27 lines, 114 (with headline 118) × 68 mm. Type: italic. Ornamental capital on 2^a. Border enclosing title. Marginal diagrams.

1^a, title; 1^b, epigram; 2^a, Iohannes Wolphius lectori . . . Norembergæ ê schola Sebaldina. 5. Calen. Septem. 1527; 3^a, rudimenta arithmetices; 28^a, Elementale Geometricum; 56^a, M. D. XXXIIII; 56^b, blank.

B.M.

VII.

1534. Elementale Geometricum. Paris. Chrestien Wechel.

ELEMEN=|TALE GEOMETRI|CVM, EX EVCLIDIS GE=|ometria, à IOANNE Vœgelin, Hayl|pronnensi, ad omnium Mathema=|tices studiosorum utilitatem | decerptum. || [DEVICE.] || PARISIIS | Apud Christianum Wechelum sub | scuto Basiliensi, Anno | M.D.XXXIIII. 36^b. [DEVICE.]

Octavo. A–D^8 E^4. 36 leaves. 8^a: 29 lines, 125 × 56 (with diagrams 80) mm. Type: italic. Marginal diagrams.

Riccardi records an edition of 1533 by the same printer.

Riccardi.

B.M.

VIII.

1536. Elementale Geometricum. Wittenberg. Joseph Klug.

ELEMENTA | GEOMETRIAE EX EVCLIDE | singulari prudentia collecta à | Ioanne Vogelin professore | Mathematico in scho⸗|la Viennensi. || ARITHMETICAE PRACTI⸗|cæ per Georgium Peur⸗|bachium Mathemati⸗|cum. | Cum præfacione Philippi Me⸗|lanthonis. 47^a. Impressum Vitebergæ per | Iosephum Klug. || M. D. XXXVI.

Octavo. A^8; B–F^8. 48 leaves, the last blank. 13^a: 26 lines, 112 × 59 (with diagram 85) mm. Types: italic, some greek. Woodcut border. Ornamental capital on 2^a. Marginal diagrams.

1^a, title within woodcut border; 1^b, blank; 2^a, P. Melanthon Ioanni Reifenstein, dated mense augusto, 1536; 8^b, epigram of Georgius Aemylius; 9^a, text; 47^a, colophon; 47^b, 48, blank.

The Arithmetic of Peurbachius forms a separate volume with separate title-page.

The words 'Philippi Me=|lanthonis' on the title-page of the copy above described have been inked out by the clerical censor, who has, however, spared the preface.

Riccardi.

Bodl.

IX.

1537. Platone Tiburtino. Johann Petreius. Nuremberg.

Continentur in hoc libro. Rudimenta astronomica Alfragani. Item Albategnius ... Ioannis de Regiomonte ... introductio in Elementa Euclidis, &c.

Quarto.

Riccardi, *Bib. mat.* ii, col. 287.

X.

1539. Elementale Geometricum, with the Elementa Arithmetices of G. Peurbachius.

Venice. J. A. de Nicolinis de Sabio, sumptu Melchioris Sessae.

ELEMENTA | GEOMETRIÆ EX EVCLIDE | singulari prudentia collecta à Ioãne Vo=|gelin professore Mathematico in | schola Viennensi. | Arithmeticæ practicæ per Georgium | Peurbuchium Mathematicum | Cum præfacione Philippi | Melanchthonis. || [DEVICE.] 32^a. ELEMENTA | ARITHMETICES. | ALGORITHMVS DE NV=| meris integris, fractis, Regulis | communibus, & de Pro=|portionibus. | Authore Georgio Peurbachio. | Omnia recens in lucem ædita fide & | diligentia singulari. | Cum præfatione Philip. Melanth. 67^b. Venetijs per Ioan. Anto. de Nicolinis de Sabio. | Sumptu uero D. Melchioris Sessæ. Anno | Domini M D XXXVIIII. | Mense Ianuario. 68^a. [DEVICE.]

Octavo. A–H^8 I^4. 68 leaves, 2[?]–67 so numbered. 8^a: 29 lines, 123 (with foliation 126) × 65 (with diagrams 83) mm. Type: italic. Marginal diagrams.

1^a, title-page, Geometry; 1^b, blank; 2^a, preface of Melancthon ?]; 7^a, epigram of G. Æmylius; 7^b, text; 32^a, title-page, Arithmetic; 32^b, epigram; 33^a, preface of Melancthon [?]; 34^a, text; 65^a, De proportionibus ex elementis Euclidis per Ioannem Vogelin; 67^b, colophon; 68^a, blank; 68^b, device.

The copy catalogued above has suffered at the hands of the clerical censor. Not only has he inked out the name of Melancthon on both title-pages, but has further cut out ff. 2–6 and 33, which presumably contain Melancthon's prefaces.

Riccardi.

B.M.

XI.

1550. Elementale Geometricum. Paris. Chrestien Wechel.

ELEMEN-|TALE GEOMETRI-|cum, ex Euclidis Geometria, | à Ioãne Vœgelin Haylprõ-|nensi, ad omnium Ma-|thematices studiosorũ | vtilitatẽ decerptũ. || [DEVICE.] || PARISIIS | Apud Christianum Wechelum, | sub Pegaso, in vico Bellouacensi. | 1550. 36^b. [DEVICE.]

Octavo. A–D⁸ E⁴. 36 leaves. Catchwords. 4ᵃ: 28 lines, 120×55 (with diagrams 78) mm. Type: italic. Marginal diagrams.

Riccardi.

Paris, Bib. Mazarine.

XII.

1551. Liber decimus elementorum. Paris. Michel Vascosan.

Euclidis elemẽtorum | LIBER DECIMVS, PETRO | Montaureo interprete. || Ad Ioannem Bellaium Cardinalem. || LVTETIÆ || Apud Vascosanum, uia Iacobæa ad insigne Fontis. || M.D.LI. || CVM PRIVILEGIO.

Quarto. a–c⁴ d⁶; B–Z AA–NN⁴. 158 leaves, 19–158 numbered 1–140. 20ᵃ: 28 lines (mixed types), 163 (with headline) 175×99 mm. Types: text roman, italic, preface roman; some greek. Ornamental capitals. Diagrams in text.

1ᵃ, title; 1ᵇ, Priuilegii sententia; 2ᵃ, praefatio; 19ᵃ, text; 157ᵃ, P. Montaurei carmen.

Riccardi.

B.M. C.T–S.

XIII.

1552. A re-issue of the preceding with title-page and preface re-set.

Euclidis elementorum || LIBER DECIMVS, PETRO || Montaureo interprete. || Ad Ioannem Bellaïum Cardinalem. || LVTETIÆ, | Apud Vascosanum, uia Iacobæa ad insigne Fontis. || M. D. LII. || CVM PRIVILEGIO.

A list of errata is placed after the preface on 18ᵇ.

Riccardi.

U.L.C.

XIV.

1551. R. Recorde. Pathway to knowledge. London. Reynold Wolfe.

The pathway to | KNOWLEDG, CONTAI-|NING THE FIRST PRIN-|ciples of Geometrie, as they | may moste aptly be applied vn-|to practise, bothe for vse of | instrumentes Geome-|tricall, and astrono-|micall and | also for proiection of plattes in euerye | kinde, and therfore much ne-|cessary for all sortes of | men. 92ᵃ. IMPRINTED at London in Poules | churcheyarde, at the signe of the Bra-|sen serpent, by Reynold wolfe. || Cum priuilegio . . . | . . . | ANNO DOMINI. M.D.LI. 92ᵇ. [DEVICE.]

Quarto. ¶ ʒ ʒ⁴; A–H⁴ I²; *a*⁴; *b–l*⁴ *m*². 92 leaves, 12 blank. 17ᵃ: 35 lines, 145 (with headline 151)×85 (with marginalia 98) mm. Types: italic, black-letter; some greek. Diagrams in text.

B.M. Bodl.

XV.

1553. Giambattista Benedetti. Resolutio omnium Euclidis problematum. Venice. Bartolommeo Caesano.

Resolutio omnium Euclidis problematum aliorumq; . . . una tantummodo circini data apertura, &c.

Quarto.

Riccardi, *Bib. mat.*, par. 1, vol. 1, col. 110.

XVI.

1555. Books 7, 8, 9 in the German translation of Scheubel. Augsburg.

Das sibend, acht und neunt Buch des hochberühmten Mathematici Euclidis Meg. in welchen der Operationen Reguln aller gemeinen Rechnung, Ursach, Grund und Fundament angezeigt wirt, zu gefallen allen den, so die kunst der Rechnung lieb haben, durch Magistrum Iohann Scheybel . . . aus dem Lateyn ins teutsch gebracht, und mit gemeinen Exemplen illustrirt, &c.

Quarto.

Murhard, ii, p. 28. Graesse.

Reprinted 1558 according to Poggendorff.

XVII.

1556. Elementorum liber 1. [With other selected propositions. Edited by V. Nabod.] Cologne. Heirs of Arnold Birckmann.

EVCLIDIS | MEGARENSIS MA-|THEMATICI CLARISS. E-|lementorum Geometricorum liber pri-|mus, cui ex sequentibus libris accesserūt | Propositiones selectæ atque ita or|dinatę, ut demonstrari queant | obseruata Geometrarum | Methodo. || [DEVICE.] || COLONIÆ, | Apud hæredes Arnoldi Birckmanni. | 1556.

Octavo. A B[8]. 16 leaves, the last blank. 4[a]: 23 lines (with spaces), 107 (with headline 113) x 61 mm. Types: text roman, preface italic; some greek. Ornamental capital on 2[a].

1[a], title; 1[b], blank; 2[a], preface of Valentinus Nabod; 3[a], text; 16, blank.

Enunciations only. Nothing is known of the editor.

B.M.

XVIII.

1559. Strasburg. Josias Rihel.

ΕΥΚΛΕΙΔΟΥ ΤΩΝ | ΠΕΝΤΕ ΚΑΙ ΔΕΚΑ ΣΤΟΙΧΕΙΩΝ | ἐκ τῶν τοῦ Θέωνος συνουσιῶν | τὸ πρῶτον. || [DEVICE.] | ARGENTORATI | Per Iosiam Rihelium. Anno | M. D. LIX.

Octavo. A–C[8] D[4]. 28 leaves, the last blank, 2–27 so numbered. 7[a]: 29 lines (mixed types), 112 (with headline 188) x 64 mm. Types: greek (2 sizes), some roman (preface), italic (title). Ornamental capital on 2[a]. Diagrams in text.

1[a], title; 1[b], blank; 2[a], Wendelinus Rihelius Ioanni Meiero, dated August, 1538; 3[b], text; 27[b], blank; 28, blank.

Riccardi.

C.T–S.

XIX.

1559. P. de La Ramée. Scholae mathematicae. Frankfort.

Petri Rami scholæ mathematicæ.

Quarto.

Murhard, ii, p. 22. 'Discursus de quindecim libris Euclidis.'

XX.

1559. Ioannes Buteo. De quadratura circuli et annotationum opuscula in errores Campani, etc. Lyons. Guillaume Rouille.

IOAN. | BVTEONIS | DE QVADRATVRA | circuli Libri duo, vbi multorum | quadraturæ confutantur, & ab | omnium impugnatione | defenditur Archi-|medes. | EIVSDEM, | annotationum opuscula in errores Campani, | Zamberti, Orontij, Peletarij, Io. Penæ | interpretum Euclidis, || [DEVICE.] || LVGDVNI, | APVD GVLIELMVM ROVILLIVM, | SUB SCVTO VENETO. | M.D.LIX. | Cum priuilegio Regis.

Octavo. a–s⁸. 144 leaves, 143, 144 blank, 3–142 paginated 5–283. 4ᵃ: 27 lines, 127 (with headline 132) × 72 mm. Types: italic; some roman and greek. Ornamental capitals. Diagrams.

1ᵃ, title; 1ᵇ, privilege; 2ᵃ, statement of contents; 3ᵃ, liber primus (secundus) de quadratura circuli; 104ᵃ, liber in errores interpretum Euclidis; 134ᵇ, annotatio in errores I. Penæ interpretis Catoptricorum; 139ᵇ, annotatio in errores I. Penæ interpretis Opticorum Euclidis; 142ᵇ, errata; 143, 144, blank.

B.M.

XXI.

1561. Elementale Geometricum. Frankfort. Heirs of C. Egenolff.

ELEMENTALE | GEOME=|METRICVM, EX EVCLIDIS GEO-|metria, à Ioanne Vœgelin, Hayl-|pronnensi, ad omnium Mathema|tices studiosorum utilita=|tem decerptum. || [DEVICE.] || Cum Gratia & Priuilegio | Cæsareo. || FRANC. Apud Hær. Chr. Ege. | Anno M. D. LXI. 35ᵇ. FRANCOFORTI | Excudebant Hæredes Christ. | Egenolphi. Anno | M.D.LXI.

Octavo. A–D⁸ E⁴. 36 leaves, the last blank. 14ᵃ: 27 lines (mixed types), 111 (with headline 117) × 65 (with diagram 83) mm. Types: roman and italic. Ornamental capital on 2ᵃ. Marginal diagrams.

B.M.

XXII.

1564. Books 1, 2 in Greek with Latin translations, &c. Edited by C. Dasypodius. Strasburg. Christian Mylius.

Part 1. *ΕΥΚΛΕΙΔΟΥ | ΤΩΝ ΠΕΝΤΕ ΚΑΙ ΔΕΚΑ ΣΤΟΙ-|ΧΕΙΩΝ, ΕΚ ΤΩΝ ΤΟΥ ΘΕΩΝΟΣ | συνουσιῶν τὸ πρῶτον.* || EVCLIDIS QVINDE-|cim Elementorum Geometriæ | primum: ex Theonis Commen-|tarijs Græcè, & Latinè. || Cui accesserunt | Scholia, in quibus quæ | ad percipienda Geometriæ Ele-|menta spectant, breuiter & dilucidè ex-|plicantur, authore Cunrado Da-|sypodio, Scholæ Argenti=|nensis professore. || ARGENTORATI EXCV-|debat Christianus Mylius, || M.D.LXIIII.

Octavo. A⁴ B–N⁸ O⁴. 104 leaves. Pagination, after 4 unnumbered leaves and 1 unnumbered page, 2–189. Catchwords. 8ᵃ: 24 lines, 111 (with headline 118) × 65 mm. Types: greek, italic, roman. Ornamental capital on 2ᵃ. Tailpieces. Diagrams in text.

1ᵃ, title; 1ᵇ, blank; 2ᵃ, Dasypodius lectori; 4ᵇ, Greek text with Latin translation on opposite page; 84ᵇ, scholia autore C. Dasypodio; 104ᵃ, errata; 104ᵇ, blank.

Part 2. ***ΕΥΚΛΕΙΔΟΥ*** | ***ΤΩΝ ΠΕΝΤΕ ΚΑΙ ΔΕΚΑ ΣΤΟΙ-|ΧΕΙΩΝ ΕΚ ΤΩΝ ΤΟΥ ΘΕΩΝΟΣ*** | *συνουσιῶν τὸ δεύτερον.* || *Καὶ* | ***ΒΑΡΛΑΑΜ ΜΟΝΑΧΟΥ ΑΡΙΘΜΗ-***|*τικὴ αποδειξις τῶν γραμμικῶς ἐν τῳ* | *δευτέρῳ τῶν ϛοιχειῶν ἀπο-*|*δειχθέντων.* || Id est, | EVCLIDIS QUINDECIM | elementorum Geometriæ secundū : ex | Theonis commentarijs | Græcè *&* Latinè. || Item, | Barlaam monachi Arithmetica demonstra-|tio eorum, quæ in secundo libro elementorum | sunt in lineis *&* figuris planis | demonstrata. || Item, | Octo propositiones Stereometricæ, eiusdem cum præ=|cedentibus argumenti. Per Cunradum Da=|sypodium, scholæ Argentinen=|sis Professorem.

Octavo. a^8 A–H^8. 72 leaves, the last blank. Pagination, after 8 unnumbered leaves and 1 unnumbered page, 2–126. Catchwords. 10^a: 23 lines, 110 (with headline 116) × 65 mm. Types: greek, italic, roman. Ornamental capital on 2^a. Tailpieces. Diagrams in text.

1^a, title; 1^b, blank; 2^a, dedication to Nicolaus Christopherus Radziwill; 8^b, Greek text with Latin translation on opposite page; 43^b, Greek text of Barlaam with Latin translation; 67^b, Theoremata octo; 72, blank.

Part 3. PROPOSITIO-|nes reliquorum Libro-|rum Geometriæ Euclidis, Græ-|cè, & Latinè, in vsum eorum, | qui || volumine Eucli-|dis carent. || Per Cunradum Dasypodium, scholæ | Argentinensis professorem. || ARGENTORATI APVD | Christianum Mylium. | M. D. LXIIII.

Octavo. a^8; A–N^8. 112 leaves. Pagination, after 8 unnumbered leaves and 1 unnumbered page, 2–205. 10^a; 22 lines, 110 (with headline 117) × 66 mm. Types: greek, italic, roman. Ornamental capital on 2^a. Tailpieces.

1^a, title; 1^b, blank; 2^a, dedication to Nicolaus Christopherus Radziwill; 8^b, Greek text, Bks. 3–13, with Latin translation; 111^b, errata; 112^b, blank.

Riccardi.

Bodl. C.T–S.

XXIII.

1564. C. Herlinus & C. Dasypodius. Demonstrationes. Strasburg.

Christiani Herlini et Conradi Dasypodii Euclideæ demonstrationes in syllogismos resolutæ.

Folio.

Murhard, ii, p. 41. Riccardi.

XXIV.

1564. Cosimo Bartoli. Modo di misurare. Venice. Francesco Franceschi.

COSIMO | BARTOLI | GENTIL' HVOMO, ET | ACCADEMICO FIORENTINO, | DEL MODO DI MISVRARE | le distantie, le superficie, i corpi, le | piante, le prouincie, le prospettiue, & | tutte le altre cose terrene, che posso-|no occorerre a gli huomini, | Secondo le uere regole d'Euclide, & de gli altri | piu lodati scrittori. || IN VENETIA, Per Francesco Franceschi Sanese. 1564.

Quarto. A^4; A–Z Aa–Nn^4. 148 leaves, 5–145 numbered 1–141. Catchwords. 6^a:

29 lines, 169 (with headline 175) x 110 mm. Types: text italic, title, dedication, chapter-headings roman. Ornamental capitals. Border on 1a enclosing title. Woodcuts (including portrait of Bartoli on 2a) and diagrams.

B.M. C.T–S.

XXV.

1564. Arithmetices Euclideae liber primus alias septimus. Wittenberg.

ARITHMETICES | EVCLIDEAE | LIBER PRIMVS. | Aliâs in ordine reliquorum | SEPTIMVS : Qui citra | præcedentium Sex librorum | Geometricorum opem eruditè | persequitur, cum reliquis duobus | sequentibus, uera principia ac | solidiora fundamenta Logi-|stices, id est, ut uocant, | Arithmetices Pra-|cticæ. || Per | Ioan. Sthen. Luneb. | IN SCHOLARVM VSVM κατὰ τὸ ὅτι | tractatus ἐρωτηματικῶς, disquisitione nimi-|rum Dialectica quæ Dialogorum | est propria. | 1564. 107b. Perscriptum Mense Iunio. | WITTENBERGÆ. | ANNO | 1564.

Octavo. A^8; B–N^8 O^4. 108 leaves. 11a: 25 lines, 117 x 63 mm. Types: roman, greek, italic. Ornamental capitals.

B.M.

XXVI.

1565. Arnold de Lens. Eisagoge in geometrica elementa. Antwerp. Christophe Plantin.

IN GEOMETRICA | ELEMENTA | EISAGOGE, | ARNOLDO DE LENS MEDICO AC | MATHEMATICO AVTORE. || *Ἡ κρίσις χαλεπή.* || [DEVICE.] || ANTVERPIÆ, | Ex officina Christophori Plantini. | ANN. M. D. LXV. | CVM PRIVILEGIO. 19b. EXCVDEBAT CHRISTOPHORVS PLAN-|TINVS, ANTVERPIÆ, IX. CALEND. | NOVEMB. ANN. M. D. LXV.

Octavo. A B^8 C^4. 20 leaves, the last blank (?). Pagination, after 1 unnumbered leaf, 3–37. 7a: 31 lines, 128 (with headline 132) x 72 (with marginalia 80) mm. Types: roman, italic, some greek. Ornamental capitals. Tailpieces. Diagrams in text.

1a, title; 1b, privilege; 2a, dedication; 2b, præfatio; 4a, text; 19b, colophon; 20, probably blank.

In his dedication the author says, 'Selegeram nuper ex Euclide præcipua quædam Theoremata, & in demonstrationibus Geometricis maximè vtilia. . . .'

Riccardi.

B.M. Bodl.

XXVII.

1567. C. Dasypodius. Volumen primum mathematicum. Strasburg. Josias Rihel.

VOLVMEN PRIMVM | MATHEMATICVM. | PRIMA, ET SIMPLICISSIMA | Mathematicarum disciplina-|rum principia com-|plectens: || GEOMETRIAE. | LOGISTICAE. | ASTRONOMIAE. || GEOGRAPHIAE. || PER CVNRADVM DASYPODIVM | in utilitatem Academiæ Argentinensis | collectum. || Vnà cum Classium eiusdem Academiæ, | ordinarijs lectionibus. || ARGENTORATI EXCVDEBAT | Iosias Rihelius, M. D. LXVII.

Octavo. A^8; A–N^8; O^8. 120 leaves, 8, 111, 112, 119, 120 blank, 9–110 numbered 1–100 with errors. 5^a: 28 lines, 117 (with headline 122) x 64 mm. Types: roman, greek, italic. Ornamental capital on 2^a.

The definitions of Bks. 1–11 of Euclid are contained in ff. 1–17 of the text.

B.M.

XXVIII.

[1568?] Jean de Merliers. L'art de mesurer toutes superficies rectilignes.
Paris. Denis du Pré.

L'ART DE | Mesurer toutes | superficies Recti-|lignes tiré des Elements | d'Euclide. || ❧ PAR IEHAN DE | Merliers d'Amiens. | [DEVICE.] | A PARIS, | Par Denis du Pré, Imprimeur, demeurant | en la rue des Amandiers, à l'enseigne | de la Verité.

Octavo. A–E^4. 20 leaves, the last blank. 4^a: 22 lines, 131 x 74 mm. Types: italic, roman. Ornamental capitals.

The dedicatory epistle of J. de Merliers to the Comte de Brissac is dated Paris, 10 Juin, 1568.

Bodl.

XXIX.

1570. Strasburg. Heirs of Christian Mylius.

Part 1. ***ΕΥΚΛΕΙΔΟΥ** | **ΣΤΟΙΧΕΙΩΝ ΤΟ ΠΡΩ-**|**ΤΟΝ.**||**ΗΡΩΝΟΣ ΑΛΕΞΑΝΔΡΕΟΣ** | ὀνόματα γεωμετρικά.* || EVCLIDIS ELEMEN=|torum Liber primus. || HERONIS ALEXANDRI-|ni vocabula geometrica: antehac | nunquam edita. græcè | *&* latinè. || Per Cunradum Dasypodium, in vsum | Academiæ Argentinensis. || ARGENTORATI | Apud hæredes Christiani Mylij. | M. D. LXX.

Part 2. ***ΕΥΚΛΕΙΔΟΥ** | **ΠΡΟΤΑ'ΣΕΙΣ** | Στοιχειῶν.* ῑe *Οπτικῶν* | *κατοπτρικῶν. Ἁρμονικῶν.* | *καὶ* | *Φαινομένων.* || EVCLIDIS | Propositiones. | Elementorum. 15. Opticorum. | Catoptricorum. Harmonicorum. | Et | Apparentiarum. || Per Cunradum Dasypodium. || ARGENTORATI | Apud hæredes Christiani Mylij. | Anno M.D. LXX.

Octavo. Pt. 1: $ɔc^8$; A–G^8; A–I^8 K^4. 140 leaves, the last blank, 10–64 numbered 2–50 with errors, 66–139 numbered 2–75. Pt. 2: a^8; b–l^8 m^4; a–l^8 m^4. 184 leaves, 8, 184 blank, 10–92 numbered 2–81 with 93–183 numbered 1–87, with errors. Pt. 1, 14^a: 24 lines, 114 (with headline 118) x 65 mm. Types: greek and italic, some roman. Ornamental capitals. Tailpieces. Diagrams in text.

Pt. 1: 1^a, title; 1^b, blank; 2^a, præfatio; 9^a, *Στοιχεῖον πρῶτον*; 46^b, *Ονόματα γεωμετρικά*; 54^a, *Ονόματα στερεωμετρικά*; 65^a, Elementum primum; 101^b, scholia C. Dasypodii; 119^b, geometriæ nomina; 128^a, stereometriæ nomina; 139^b, blank; 140, blank. Pt. 2: 1^a, title; 1^b, blank; 2^a, præfatio; 7^b, ad lectorem; 8, blank; 9^a, *Στοιχεῖα*; 61^a, *Ὀπτικὰ καὶ κατοπτρικά*; 73^a, *Ἐισαγωγὴ ἁρμονική*; 85^b, *Κατατομὴ κανόνος*; 87^b, *Φαινόμενα*; 93^a, Geometriæ libri xv; 148^a, Optica; 157^a, Catoptrica; 160^a, Musica 176^b, Phænomena; 183^a, colophon; 183^b, blank; 184, blank.

Riccardi.

B.M. Bodl.

XXX.

1571. Strasburg. Christian Mylius.

Part 1. EVCLIDIS | ELEMENTORVM | Liber primus. | Item, | HERONIS ALEXANDRINI | vocabula quædam geometrica: ante hac nun=|quam edita, græcè & latinè. | Per M. Cunradum Dasypodium. | [DEVICE.] | Cum gratia & priuilegio Cæsareo, atque | Regis Gallię, ad sexennium. || ARGENTINÆ, | 1571.

Part 2. EVCLIDIS | OMNES OMNIVM LI=|BRORVM PROPOSITIO=|nes, græcè & latinè: | editæ per || M. Cunradum Dasypodium. | [DEVICE.] | Cum gratia & priuilegio Cæsareo, | Regisq; Galliæ, ad sexennium. || ARGENTINÆ, | 1571 207^b. [DEVICE.] | ARGENTORATI APVD | Christianum Mylium. | 1571.

Octavo. Pt. 1: as preceding edition. Pt. 2: a^8; b–n^8; a–n^8. 208 leaves, 104, 208 blank, 10–103 numbered 2–93, 105–207 numbered 1–98, with errors.

Pt. 1: Contents as preceding edition. Pt. 2: 1^a, title; 1^b, Opera Euclidis; 2^a, præfatio; 7^b, index; 8^b, blank; 9^a, *Στοιχεῖα*; 61^a, *'Οπτικὰ καὶ κατοπτρικά*; 73^a, *'Εισαγωγὴ ἁρμονική*; 85^b, *κατατομὴ κανόνος*; 87^b, *Φαινόμενα*; 93^a, *Δεδομένα*; 104, blank; 105^a, Geometriæ libri xv; 160^a, Optica; 169^a, Catoptrica; 172^a, Musica; 188^b, Phænomena; 195^a, Data; 207^b, device and colophon; 208, blank.

A very close reprint of the preceding edition, many of the sheets in Pt. 1 (the preface excepted) appearing to be identical. Pt. 2 is also a close reprint but has additional matter, viz. indexes on 1^b, 7^b, and 8^a and the *Δεδομένα* with Latin translation which had been promised in the note to the reader on 7^a of Pt. 2 of the 1570 edition.

Riccardi.

B.M. Bodl.

XXXI.

1574. Robert Recorde. Pathway to knowledge. John Harrison. London.

¶ The pathewaie to | knowledge, containyng the first | principles of Geometrie, as thei | maie moste aptly bee applied vnto | practise, bothe for vse of instru=|mentes Geometricall, and A-|stronomicall: and also for | proiection of plattes | in euery kinde, | and | therefore muche neces=|sarie for all sortes | of menne. || Geometries verdicte. [an epigram] || 1574. 92^a. IMPRINTED AT LON-|don, by Ihon Harrison. | Anno Domini. | 1574.

Quarto. *, $*^4$; A–H^4 I^2; a^4; A^4 b–l^4 m^2. 92 leaves, 8 blank. 13^a: 35 lines, 150 (with headline 159) × 90 (with diagrams 109) mm. Types: black-letter; some roman and greek. Ornamental capitals. Diagrams in text.

1^a, Bk. 1, title; 1^b, Argumentes of the fower bookes; 2^a, To the gentle reader; 3^a, Dedication to King Edward VI; 8, blank; 9, Conclusions Geometricall; 42^b, blank; 43^a, Bk. 2, title; 43^b, epigram; 44^a, Preface unto the Theoremes; 46^b, blank; 47^a, Preface declaryng briefly the commodities of Geometrie; 51^a, Theoremes Geometricall; 92^a, colophon; 92^b, blank.

B.M. (imp.) Bodl.

XXXII.

1575. Definitiones elementi quinti et sexti ab Elia Vineto Santone interpretatae. Bordeaux. Simon Millanges.

DEFINITIONES | ELEMENTI QVINCTI, | ET SEXTI, EVCLIDIS, AB | ELIA

VINETO SANTONE | INTERPRETATÆ. | ❧ || [DEVICE.] || BVRDIGALÆ, | Apud Simonem Millangium Burdigalensium | Typographum, via Iacobea, | 1575.

Quarto. A–E⁴ F⁶. 26 leaves. 4ᵃ: 33 lines, 156 (with headline 161) × 99 (with marginalia 108) mm. Types: roman, italic, some greek. Ornamental capital on 2ᵃ. Head- and tailpieces. Diagrams in text.

1ᵃ, title; 1ᵇ, quotation from Cicero; 2ᵃ, Elias Vinetus lectori; 3ᵃ, text, Bk. v; 19ᵃ, text, Bk. vi; 25ᵇ, index; 26ᵇ, blank.

The preface of Vinetus is dated Montiniaci ad Carántonum . . . M.D.LIX.

Riccardi.

Bodl.

XXXIII.

1575. Francesco Maurolico. Opuscula Mathematica.
Venice. Francesco Franceschi.

D. FRANCISCI | MAVROLYCI, | ABBATIS MESSANENSIS, | Opuscula Mathematica; | Nunc primum in lucem ædita, cum rerum omnium | notatu dignarum. | INDICE LOCVPLETISSIMO. | PAGELLA HVIC PROXIME CONTIGVA, | eorum Catalogus est. | [DEVICE.] | CVM PRIVILEGIO. | Venetijs, Apud Franciscum Franciscium Senensem. | M D L XXV. 155ᵃ. D. FRANCISCI | MAVROLYCI, | ABBATIS MESSANENSIS, | Mathematici celeberrimi, | ARITHMETICORVM LIBRI DVO, | NVNC PRIMVM IN LVCEM EDITI, | Cum rerum omnium notabilium. || INDICE COPIOSISSIMO. | [DEVICE.] CVM PRIVILEGIO. | Venetijs, Apud Franciscum Franciscium Senensem. | M D LX XV.

Quarto. a¹⁰; A–S⁸; T⁴ V–Z Aa–Ff⁸ Gg¹⁰ Hh⁸. 256 leaves, 154, 256 blank. 13ᵃ: 41 lines, 171 (with headline 176) × 106 mm. Types: text roman, chapter- and section-headings italic. Ornamental capitals. Diagrams in text and margin.

1ᵃ, title; 1ᵇ, catalogus; 2ᵃ, dedication; 3ᵃ, index; 11ᵃ, De sphæra; 23ᵇ, Computus ecclesiasticus; 34ᵇ, Tractatus instrumentorum; 50ᵇ, De lineis horariis; 62ᵃ, Euclidis Elementorum liber tredecimus, solidorum tertius, & Regularium corporum primus; 83ᵃ, Musicae traditiones; 91ᵃ, De lineis horariis; 153ᵇ, blank; 154, blank; 155ᵃ, second title; 155ᵇ, blank; 156ᵃ, table and formatio numerorum; 159ᵃ, Arithmetica; 248ᵇ, device; 249ᵃ, index lucubrationum; 251ᵃ, index; 256, blank.

Riccardi.

B.M. Bodl.

XXXIV.

1579. Isaaci monachi scholia. Translated by C. Dasypodius.
Strasburg. Nicolas Wyriot.

Isaaci Monachi | SCHOLIA IN | EVCLIDIS ELEMENTORVM | Geometriæ, sex priores | libros. | Per | CVNRADVM DASYPODIVM | in Latinum sermonem translata: | & in lucem edita. | [DEVICE.] | ARGENTORATI | Excudebat Nicholaus Vvyriot. | M. D. LXXIX.

Octavo. *⁸; A–F⁸. 56 leaves. 11ᵃ: 27 lines, 143 (with headline 148) × 68 mm. Types: italic, roman. Ornamental capitals.

Riccardi.

Bodl.

XXXV.

1583. Iohann Thomas Freigius. Quaestiones geometricae et stereometricae. Basle. Sebastian Henricpetri.

IOAN. THOMÆ | FREIGII | QVAESTIO⸗|NES GEOMETRICÆ ET | STEREOMETRICAE | in Euclidis & Rami ϛοιχέιω-|σιν: in quibus || Logica veterum Mathematicorum illustratur & demon-|stratur. || Πᾶς αγεωμέτρητος εἰσίτω. || [DEVICE.] || Cum Gratia & Priuilegio Cæs. Maiestat. || BASILEÆ, | PER SEBASTIANVM | HENRICPETRI. 192ᵇ. BASILEÆ, | PER SEBASTIANVM | HENRICPETRI, ANNO SA=|lutis CIↃ. IↃ. LXXXIII. | Mense Martio.

Octavo. α–γ^8 A–X^8. 192 leaves, the last blank, 25–191 with pagination 1–333. 18^a: 35 lines, 120 (with headline 124) x 68 mm. Types: roman, italic; some greek. Ornamental capitals.

1^a, title; 1^b, quotation from Plutarch; 2^a: dedicatory letter to C. Dasypodius and others; 4^a, procemium mathematicum; 23^b, tabula geometriæ generalis; 25^a: quæstiones geometriæ; 151^b, quæstiones stereometriæ; 188^a, letter of Lazarus Schonerus to Freigius dated 7 June, 1577; 190^a, letter of Ramus; 191^b, colophon; 192, blank.

B.M. Bodl.

XXXVI.

1589. Cosimo Bartoli. Modo di misurare. Venice. Francesco Franceschi.

COSIMO BARTOLI | GENTIL'HVOMO, ET | Accademico Fiorentino, | DEL MODO DI MISVRARE | le distantie, le superficie, i corpi, le | piante, | le prouincie, le prospettiue, & | tutte le altre cose terrene, che posso-|no occorrere a gli huomini, | Secondo le vere regole d'Euclide, & de gli altri | piu lodati scrittori. || IN VENETIA, Per Francesco Franceschi Sanese. 1589.

Quarto. A–S^8 T^4. 148 leaves, 4–145 so numbered. 6^a: 29 lines, 168 (with headline 175) x 112 mm. Types: text italic, title, dedication, chapter-headings roman. Ornamental capitals. Border on 1^a enclosing title. Woodcuts and diagrams.

A close reprint of the 1564 edition with the same border, woodcuts, and diagrams.

B.M. C.T–S.

XXXVII.

1599. Giovanni Pomodoro. Geometria prattica. Rome. Stefano de Paulini.

GEOMETRIA | PRATTICA | TRATTA Dagl' Elementi d' Euclide | et altri Auttori | Da Giouanni Pomo doro Venetiano | Mathematico eccellentissimo descritta | et Dichiarata da Giouanni Scala Mathematico. || Nella quale si uede in ·50· Tauole di Rame scol⸗|pito tutto quello che ad un buon Giometra s' ap⸗|partiene di sapere et porre in uso. || Opera non meno Vttile che necessaria, a' Misuratori di | terreni, di fabriche, et altri simmili, ma' in' oltre | ancora a', Geografi, Cosmografi, Architetti Ciuili, et | Militari, a' Bombardieri, Soldati priuati, a' | Capitani, Mastri di Campo, et a' qual si Voglia | altra persona Virtuosa. [Arms of Cardinal Aldobrandino.] || Apresso Stefano de Paulini In Roma con licenza de Superiori || 1599.

Folio. *²; A–M⁴ N⁶⁺¹. 57 leaves. Double columns. 8ᵃ: 67 lines, 280 (with headline 289) x 184 mm. Types: roman. Ornamental capitals. Tailpieces. Engraved title-page and diagrams.

Riccardi.

B.M.

XXXVIII.

1600. Petrus Ryff. Quaestiones Geometricae & Geodaesia.
Heirs of A. Wechel, Claude Marne, Iohann Aubry. Frankfort.

QUÆSTIONES | GEOMETRICÆ, | in | EuCLIDIS & P. RAMI | *ΣΤΟΙΧΕΙΩΣΙΝ*; | In usum Scholæ Mathematicæ collectæ, | a | Doctore PETRO RYFF, Basil. | Mathematum Professore. | Quibus | GEODÆSIAM | adjecimus | per usum RADII Geometrici. || [DEVICE.] || FRANCOFVRTI, | Apud Andr. Wecheli heredes, Claud. Mar-|nium, & Ioannem Aubrium. | M. DC.

Quarto. A^4; B–R^4. 68 leaves, 2–68 with pagination 3–135. 45^a: 32 lines (mixed types), 148 (with headline 153) x 99 mm. Types: roman, italic; some greek. Ornamental capitals. Woodcuts and diagrams.

1^a, title; 1^b, epigram of L. Lucius; 2^a, dedication to Andreas Ryff; 3^a, Geometriæ *Διατύπωσις*; 5^a, Quæstiones Geometricæ; 55^b, Geodæsia; 68^b, device.

B.M.

INDEX OF PLACES

(*References refer to the numbers.*)

INDEX OF PRINTERS AND PUBLISHERS

INDEX OF PRINTERS AND PUBLISHERS

PLATES

LIST OF PLATES

¶ Erhardus ratdolt Augustensis impressor. Serenissimo alme vrbis venete Principi Joanni Mocenico. S.

Solebam antea serenissime princeps mecum ipse cogitans admirari quid cause esset ꝙ in hac tua prepotenti ⁊ fausta vrbe cum varia auctorum veterum nouorumq; volumina quottidie imprimerent. In hac mathematica facultate vel reliquarum disciplinarum nobilissima aut nihil aut parua quedam et friuola in tanta impressorū copia qui in tua vrbe agunt: viderentur impressa. Hec cum mecum sepius discuterem inueniebam id difficultate operis accidisse. Non enim adhuc quo pacto schemata geometrica: quibus mathematica volumina scatent: ac sine quibus nihil ī his disciplinis fere intelligi optime potest excogitauerant. Itaq; cum hoc ipsum tantūmodo cōmuni omnium vtilitati que ex his percipitur. obstaret mea industria nō sine maximo labore effeci. vt qua⁹ facilitate litterarum elementa imprimuntur .ea etiam geometrice figure conficerentur. Quamobrem vt spero hoc nostro inuento he discipline quas mathemata greci appellant voluminum copia sicuti reliqne scientie breui illustrabuntur. De quarum laudibus ⁊ vtilitate possem multa im presens adducere ab illustribus collecta auctoribus: nisi studiosis iam omnibus hec nota essēt. Illud etiam plane cognitum est ceteraꝫ scientias sine mathematibus imperfectas ac veluti mancas eē. Neq; hoc profecto negabunt Dialectici neq; Philosophi abnuent: in quoꝝ libris multa reperiuntur: que sine mathematica ratione minime intelligi possunt. Quam diuin⁹ ille Plato mere veritatis arcanū. vt adipisceretur cyrenas ad . Theodorum summum eo tempore mathematicuꝫ ⁊ ad egiptios sacerdotes enauigauit. Quid ꝙ sine hac vna facultate viuendi ratio nō perfecte constat. Nam vt de musice taceam: que nobis muneri ab ipsa natura ad perferendos facilius labores concessa videtur: vt astrologiā preteream qua exculti celum ipsum veluti scalis machinisq; quibusdam conscendentes verum ipsius nature argumentum cognoscimus: sine arithmetica ⁊ geometria: quarum altera numeros altera mēsuras docet ciuiliter: comodeq; viuere q̄ possum⁹: Sed quid ego ī his moror que iam omnibus vt dixi: notiora sunt q̄ȝ vt a me dicantur. Euclides igitur megarensis serenissime princeps qui. xv. libris omnem geometrie rationem consummatissime complexus est: quem ego summa ⁊ cura ⁊ diligentia nullo pretermisso schemate imprimendum curaui: sub tuo numine tutus felixq; prodeat.

Preclarissimus liber elementorum Euclidis perspicacissimi: in artem Geometrie incipit quãfoelicissime:

De principijs ꝑ se notis: ⁊ p̄mo de diffinitionibus earundem.

Punctus est cuius ps nõ est. ¶ Linea est lõgitudo sine latitudine cuiꝰ quidẽ extremitates sũt duo pũcta. ¶ Linea recta ẽ ab vno pũcto ad aliũ breuissima extẽsio i extremitates suas vtrũqꝫ eoꝝ recipiens. ¶ Supficies ẽ q̃ lõgitudinẽ ⁊ latitudinẽ tm̃ hꝫ: cuiꝰ termi quidẽ sũt linee. ¶ Supficies plana ẽ ab vna linea ad aliã extẽsio i extremitates suas recipiẽs ¶ Angulus planus ẽ duarũ linearũ alternus ꝯtactus: quaꝝ expãsio ẽ sup supficiẽ applicatioqꝫ nõ directa. ¶ Quãdo aũt angulum ꝯtinẽt due linee recte rectilineꝰ angulus noiat̃. ¶ Qñ recta linea sup rectã steterit duoqꝫ anguli vtrobiqꝫ fuerit eq̃les: eoꝝ vterqꝫ rectꝰ erit ¶ Lineaqꝫ linee supstãs ei cui supstat ppendicularis vocat̃. ¶ Angulus vo qui recto maior ẽ obtusus dicit̃. ¶ Angulꝰ vo minor recto acutꝰ appellat̃. ¶ Terminꝰ ẽ qd vniuscuiusqꝫ finis ẽ. ¶ Figura ẽ q̃ tmino vl termis ꝯtinet̃. ¶ Circulꝰ ẽ figura plana vna q̃dem linea ꝯtẽta: q̃ circũferentia noiat̃: in cuiꝰ medio pũctꝰ ẽ: a quo oẽs linee recte ad circũferẽtiã exeũtes sibiinuicẽ sũt equales. Et hic quidẽ pũctꝰ cẽtrũ circuli dr̃. ¶ Diameter circuli ẽ linea recta que sup eiꝰ centꝝ trãsiens extremitatesqꝫ suas circũferẽtie applicans circulũ i duo media diuidit. ¶ Semicirculus ẽ figura plana diametro circuli ⁊ medietate circũferentie ꝯtenta. ¶ Portio circuli ẽ figura plana recta linea ⁊ parte circũferẽtie ꝯtẽta: semicirculo quidẽ aut maior aut minor. ¶ Rectilinee figure sũt q̃ rectis lineis cõtinent̃ quarũ quedã trilatere q̃ tribꝰ rectis lineis: quedã quadrilatere q̃ q̃tuor rectis lineis. q̃dã mltilatere que pluribus qꝫ quatuor rectis lineis continent̃. ¶ Figurarũ trilaterarũ: alia est triangulus hñs tria latera equalia. Alia triangulus duo hñs eq̃lia latera. Alia triangulus triũ inequalium laterũ. Haꝝ iterũ alia est orthogoniũ: vnũ .s. rectum angulum habens. Alia ẽ ambligonium aliquem obtusum angulum habens. Alia est oxigonium: in qua tres anguli sunt acuti. ¶ Figurarũ autẽ quadrilateraꝝ Alia est q̃dratum quod est equilaterũ atqꝫ rectangulũ. Alia est tetragonꝰ longꝰ: q̃ est figura rectangula: sed equilatera non est. Alia est helmuaym: que est equilatera: sed rectangula non est.

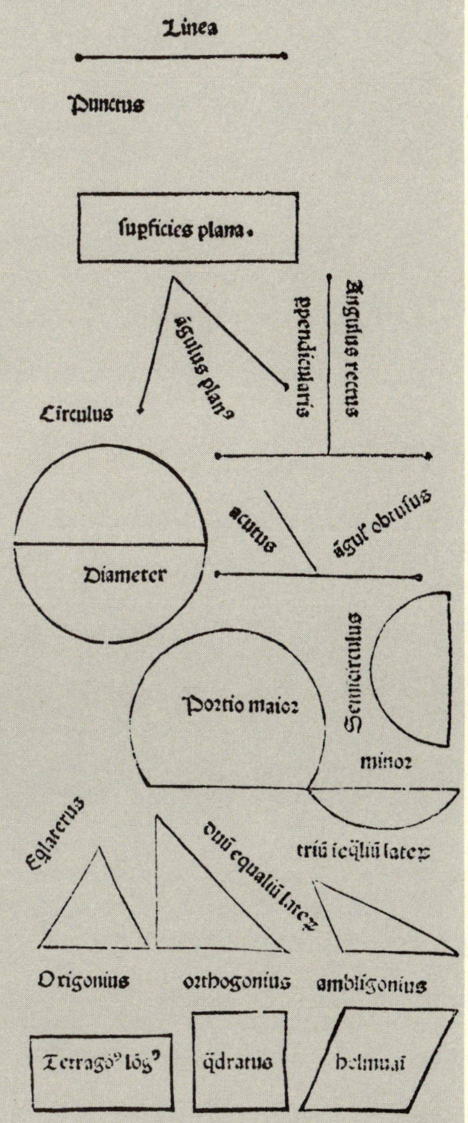

LIBER

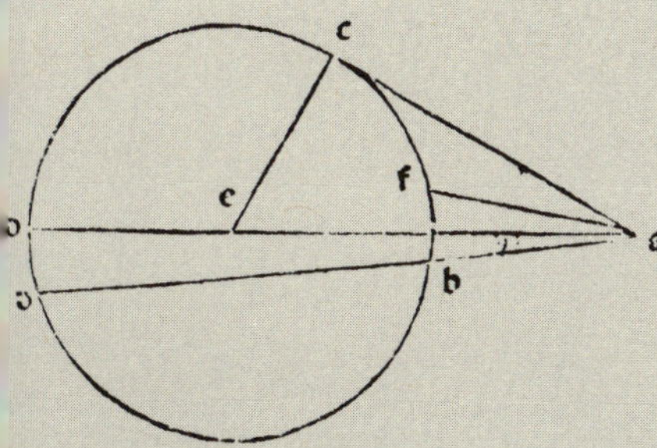

Propositio .36.

Si fuerit punctus extra circulũ signatus aquo due linee ad circũferẽtiã ducant̃ .altera secans.altera circumferentie applicata fueritq̃ʒ qḋ ex ductu totius secantis in partẽ sui extrinsecam equum ei quod ex ductu applicate in cipsam fit: erit linea applicata ex necessitate circulũ contingens.

¶ Sit a punctus signatus extra circulum.b.c.d.cuius centrum.e. a quo ducantur ad circulum linea.a.b.d.secans ipsum ⁊ linea.a.c.applicata circumferentie.⁊ esto vt qḋ fit ex.d.a.in.a.b.fit equale quadrato.a.c.dico lineam.a.c.esse contingentẽ ⁊ est hec conuersa prioris. Si enim non est contingens. sit ergo contingens linea a.f: eritq̃ʒ per premissam quod fit ex.d.a.in.a.b.equale quadrato linee.a.f.quare quadratum linee.a.f.est equale quadrato linee.a.c.ergo.a.c.est equalis.a.f.quod est impossibile.per.8.huius.erit ergo.a.c.contingens quod est propositũ. ¶ Idẽ ostẽsiue ꝓbabit̃ mancat prior dispositio ⁊ ypothesis.⁊ si linea.a.b.d.trãsit p cẽtꝝ ducatur linea.c.e.q̃ʒ erit per.6.secundi qḋ fit ex.d.a.in.a.b.cum quadrato.e.b .et ideo cum quadrato.e.c.equale quadrato.a.e.sed qḋ fit ex.d.a.in.a.b.positum est equale quadrato.a.c.ergo quadratum.a.c.cũ quadrato .c.e.ẽ equale quadrato.a e. ergo per vltimã primi angulus.c.est rectus.ergo per coroľ 15.huius linea. a.c. est contingens circulum quod est propositum. ¶ Si autem.a.b.d.non transit per centrum ducatur a puncto.d.linea transiens per centrum.⁊ quia qḋ fit ex hac to/ta in eius partem extrinsecam est equale ei quod fit ex.d.a.in.a.b.per premissam ipsum erit equale quadrato linee.a.c.quare vt prius.a.c.erit contingens circulũ.

Explicit liber tertius. Incipit liber quartus.

Figura intra figuram dicitur iscribi quando ea que inscribitur eius in qua inscribitur.latera vno quoq̃ʒ suorum angulorum ab interiore parte contingit. ¶ Circumscribi vero figura figure perhibetur quotiẽs ea quidẽ figura eius cui circũscribitur suis lateribus omnibus omnes angulos contingit.

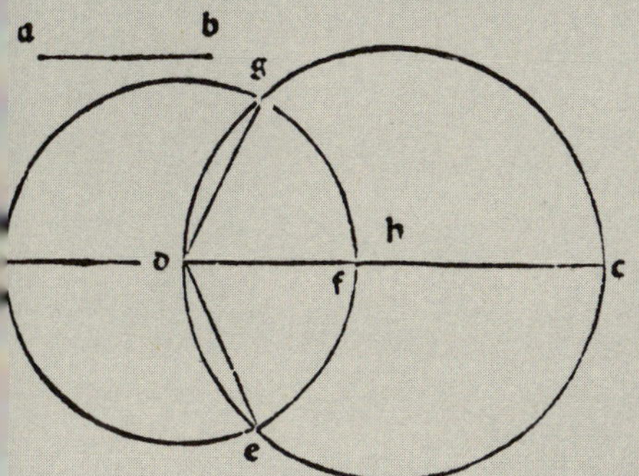

Propositio. .1.

Intra datum circulum date linee recte que diametro mi/nime maior existat equam rectam lineam coaptare.

¶ Sit linea data.a.b.circulusq̃ʒ datus.c.d. e.cui⁹ diameter.c.d.qua nõ ẽ maior linea.a.b.volo itra datũ circulũ coaptare lineã eq̃lẽ.a.b que si fuerit equalis diametro cõstat ꝓpositũ.si aũt minor ex diametro sumat̃.d.f.sibi eq̃lis ⁊ sup punctũ.d.ẽm quãtitatẽ linee.d.f.describat̃ circulus f.e.g.secans datum circulum in punctis.g.⁊.e.ad alterum quorum ducatur linea a puncto.d.vt.d.e.vel.d.g.eritq̃ʒ vtralibet earum equalis linee.a. b.eo ꝙ vtraq̃ʒ earũ est equalis linee.d.f.per diffinitionem circuli:quare habemus propositũ.

Preclarissimus liber elementorum Euclidis perspicacissimi: in artem Geometrie incipit quā foelicissime:

Punctus est cui' ps nō est. ¶ Linea est lōgitudo sine latitudine cui' quidē extremitates sūt duo pūcta. ¶ Linea recta ē ab uno pūcto ad aliū breuissima extēsio i extremitates suas utrūq3 eorū recipiēs. ¶ Supficies ē que lōgitudinē & latitudinē tantū h3: cuius termini quidē sūt linee. ¶ Supficies plana ē ab una linea ad aliā extēsio i extremitates suas recipiens ¶ Angulus planus ē duarū linearū alternus cōtactus: quarum expāsio ē sup supficiē applicatioq3 nō directa. ¶ Qñ do aūt angulum continēt due linee recte rectilineus angul' noiat. ¶ Quādo recta linea sup rectā steterit duoq3 anguli utrobiq3 fuerit eq̄les: eoꝝ uterq3 rect' erit. ¶ Lineaq3 linee superstās ei cui supstat ppendicularis uocat. ¶ Angulus uero qui recto maiorē obtusus dicit. ¶ Angulus uero minor recto acut' appellat. ¶ Termin' ē qd' uniuscuiusq3 finis est. ¶ Figura ē q̄ termio uel termis cōtinet. ¶ Circul' est figura plana una qdē linea cōtēta: q̄ circūferētia noiat: i cui' medio pūct' est: a quo oēs linee recte ad circūferētiā exeūtes sibiinuicē sūt equales. Et hic qdē pūct' centrū circuli dicit. ¶ Diameter circuli ē linea recta que sup ei' centꝝ trāsiēs extremitatesq3 suas circūferētie applicans circulū i duo media diuidit. ¶ Semicirculus ē figura plana diametro circuli & medietate circūferētie cōtēta. ¶ Portio circuli ē figura plana recta linea & parte circūferētie cōtēta: semicirculo quidē aut maior aut minor. ¶ Rectilinee figure sūt q̄ rectis lineis cōtinent quaꝝ quedā trilatere que trib' rectis lineis: quedā quadrilatere que q̄tuor rectis lineis. quedā multilatere que pluribus q̄ quatuor rectis lineis cōtinent. ¶ Figuraꝝ trilateraꝝ: alia ē triāgul' hñs tria latera equalia. Alia triāgulus duo habens eq̄lia latera. Alia triāgulus triū inequaliū laterū. Haꝝ iteꝝ alia est orthogoniū: unū. s. rectū angulū habēs. Alia ē ambligonium aliquē obtusum angulū habens. Alia ē oxigonium: i qua tres anguli sūt acuti. ¶ Figuraꝝ aūt q̄drilateraꝝ Alia ē q̄dratum quod est eqlaterū atq3 rectangulū. Alia est tetragon' long': que ē figura rectāgula: sed eqlatera nō est. Alia est helmuaym: que ē equilatera: sed rectāgula non est.

De principiis p se notis: & pmo de diffinitionibus earundem.

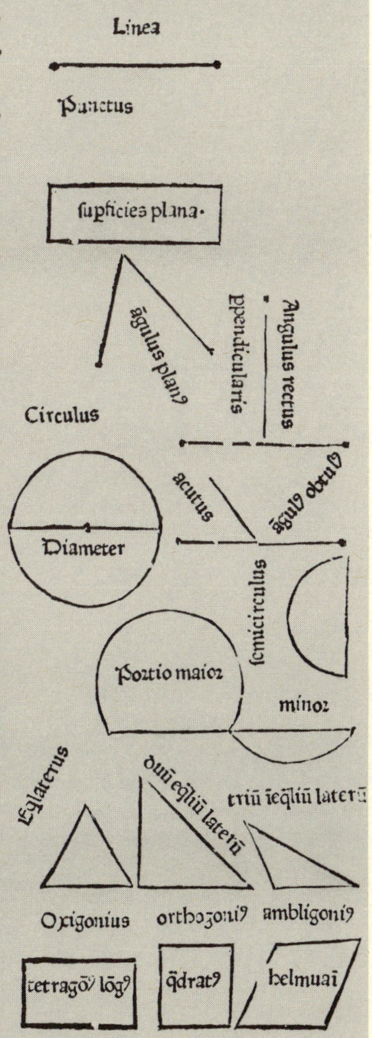

¶Euclidis Megarensis acutissimi mathematici elemen toꝝ liber primus ex traditione Theonis Bartholameo Zāberto Vene. interp̄te icipit aue foelici.

¶Diffinitio prima.

Signum est cuius pars nulla.

¶Diffinitio. ii.

L inea uero longitudo illatabilis.

¶Diffinitio. iii.

L ineæ autem limites sunt signa.

Diffinitio. iiii.

R ecta linea ē q̄ ex æq̄li sua iteriacet signa.

¶Diffinitio. v.

S uperficies est quę longitudinem: latitudinemque tantum habet.

¶Diffinitio. vi.

S uperficiei extrema sunt lineæ.

¶Diffinitio. vii.

P lana superficies est quæ ex æquali suas interiacet lineas.

¶Diffinitio. viii.

P lanus angulus ē: duarū linearum in plano se se tangentium: & non in dirrecto iacentium ad alterutram inclinatio.

¶Diffinitio. ix.

Q uando autem quę angulū continent lineæ rectæ fuerint: recti lineus angulus nuncupatur.

¶Diffinitio. x.

C um uero recta linea super rectam consistens lineam utrobiq; angulos æquales ad inuicem fecerit: rectus est uterq; æqualium angulo꞉ & quę supstat recta linea perpēdicularis uocitať sup q̄ steterit.

¶Diffinitio. xi.

O btusus angulus maior est recto.

¶Diffinitio. xii.

A ccutus uero minor est recto.

¶Diffinitio. xiii.

T erminus est quod cuiusq; finis est.

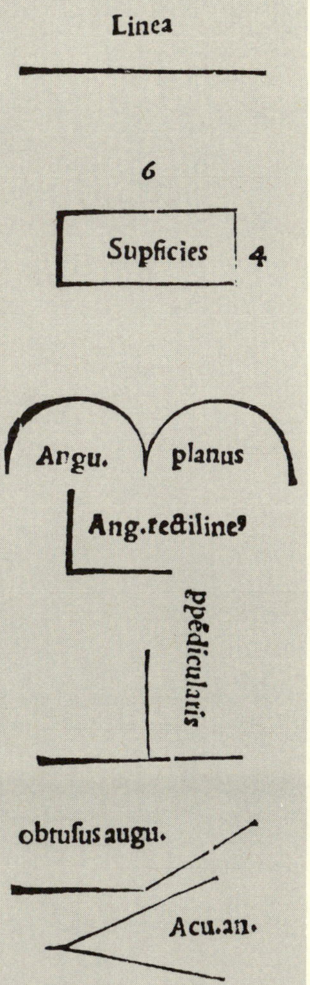

Euclidis

megarensis philo
sophi acutissimi mathematicorumqȝ omni
um sine controuersia principis opa a Cam
pano interprete fidissimo tralata Que cum
antea librariorum detestanda culpa mēdis
fedissimis adeo deformia eēnt: vt vix Eu
clidem ipsum agnosceremus. Lucas pacio
lus theologus insignis: altissima Mathe
maticaꝝ disciplinarum scientia rarissimus
iudicio castigatissimo detersit: emendauit.
Figuras cētum ⁊ vndetriginta que in alijs
codicibus inuerse ⁊ deformate erant: ad re
ctam symmetriam concinauit: ⁊ multas ne
cessarias addidit. Eundē quoqȝ plurimis
locis intellectu difficilem cōmentario
lis sane luculentis ⁊ eruditiss. ape
ruit: enarrauit: illustrauit Adhec
vt elimatior exiret Scipio ve
gius mediol. vir vtraqȝ
ligua: arte medica: subli
mioribusqȝ studijs
clarissimus diligē
tiam: ⁊ censurā
suā prestitit.

A. Paganius Paganinus Characteri-
bus elegantissimis accuratissi-
me imprimebat.

LIBER

De figurarum unius alteri inscriptione & circumscriptio ne Liber quartus. Euclidis ex suprema Campani iterpretatio ne. Magistro Luca paciolo de burgo. Sancti Sepulcri Ordi nis minorum castigatore feruentissimo. Incipit.

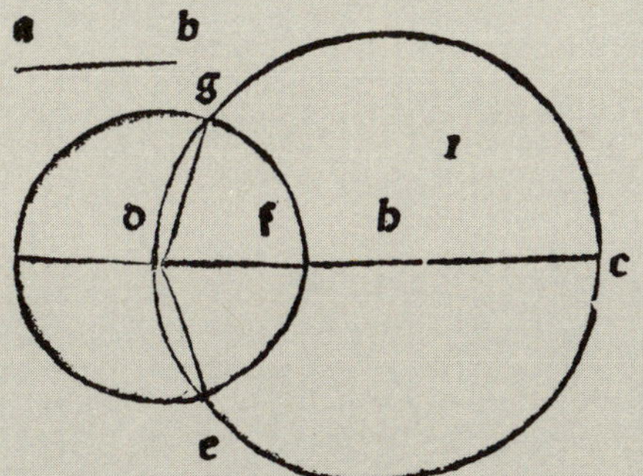

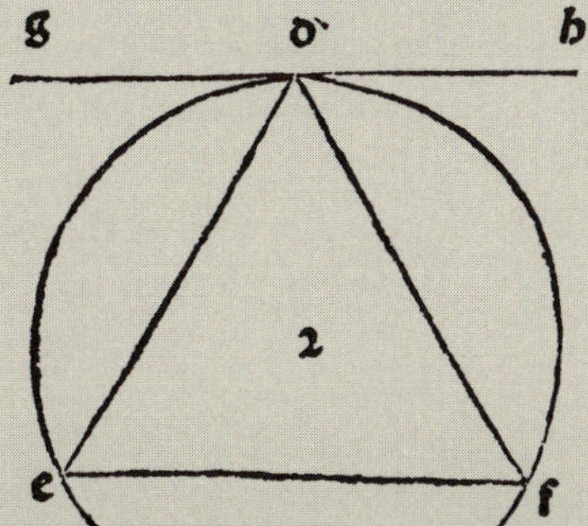

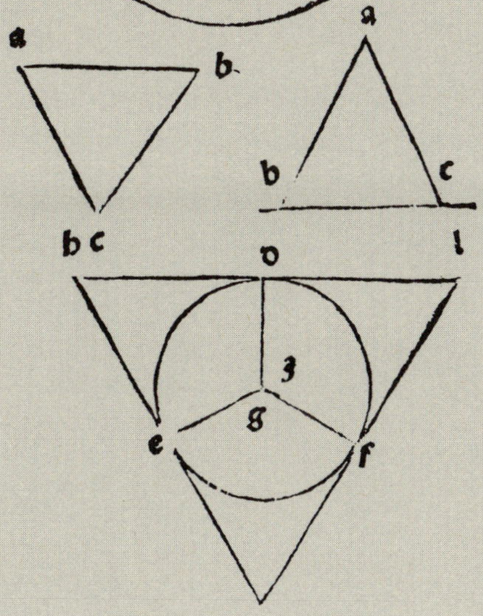

Figura intra figuram dicit' inscri bi quãdo ea que inscribitur eius in qua inscribit'. latera vno quoq; suorum angulorum ab interiore parte contingit. Circumscribi vero figura figure perhibet' quo tiens ea quidem figura eius cui circumscribit' suis lateribus om nibus omnes angulos cõtingit.

Propositio .1.

Intra datum circulu; date linee recte que diame tro minime maior existat equam rectam lineam coaptare.

Sit linea data.a.b.circulusq; datus.c.d.e.cui9diameter c.d.qua non est maior linea.a.b.uolo intra datum circu lum coaptare lineam equalem.a.b.q̃ si fuerit equalis dia metro constat propositum. Si ãt minor, ex diametro sumatur.d.f.sibi equalis.& super punctum.d.p̃m quantitatem linee.d.f.describatur circu lus.f.e.g.secans datum circulum in punctis.g.&.e.ad alte; quorum duca tur linea a puncto.d.ut.d.e.uel.d.g.eritq; utralibet earum equalis linee a.b.eo q̃ utraq; earum est equalis linee.d.f.per diffinitionem circuli: qua re habemus propositum.

Propositio .2.

Intra assignatum circulum triangulum triangu lo assignato equiangulum collocare.

Sit assignatus triangulus.a.b.c.assignatusq; circulus.d.
e.f.uolo intra hunc circulum collocare unum triãgulum
equiangulum triangulo.a.b.c.equilaterum enim non est
necessarium esse sed est possibile. Produco.g.d.h.cõtin
gentem circulum in puncto.d.super quem facio angulum.h.d.f.ducta li
nea.d.f.equalem angulo.c.& angulum.g.d.e.ducta linea.d.e.equalem
angulo.b.& protraho lineam.e.f.eritq; per.31.tertii angulus.e.equalis an 31
gulo.c.quia uterq; est equalis angulo.h.d.f.c.quidem per positionem.e.
uero per.31.tertii eadem ratione erit angulus.f.equalis angulo.b.quare p̃ 31
2.primi.d.tertius erit equalis.a.tertio.quare habemus propositum. 32

Propositio. .3.

Circa assignatum circulum assignato triangulo tri angulum equiangulum describere

Sint ut prius assignatus triangulus.a.b.c.assignatusq;
circulus.d.e.f.cuius centrum.g.circa hunc circulum uolo
describere unum triangulum eqangulum triangulo.a.b.c.
equilaterum enim non est necessarium sed ẽ possibile. Pro
ducam basim.b.c.in utranq; p̃tem.ut fiant duo anguli extrinseci.& a cen
tro.g.Producam lineam.g.d.ad circumferentiam.& constituam angu
lum.d.g.e.ducta linea.g.e.equalẽ angulo.b.extrinseco.&.d.g.f.ducta li
nea.g.f.equalẽ.c.extrinseco.& a punctis.d.e.f.producam in utranq; partẽ
lineas orthogonaliter que per correlarium.15.tertii erũt contingentes cir 15

ORONTII

FINAEI DELPHINATIS,
REGII MATHEMATI-
CARVM LVTETIAE
PROFESSORIS,

In ſex priores libros Geometricorum
elementorum Euclidis Megarēſis demonſtratio-
nes, Recèns auctæ, & emendatæ: vnà cum ipſius
Euclidis textu græco, & interpretatione
latina Bartholamæi Zamberti Ve-
neti. Omnia ad fidem geome-
tricam, per eundē Oron-
tium recognita.

LVTETIAE PARISIORVM,
Apud Simonem Colinæum.

1 5 4 4.

Cum priuilegio Regis.

Vireſcit vulnere virtus.

Die Sechs Erste Bücher

Euclidis/

Vom anfang oder grund der Geometrj.

In welchen der rechte grund/nitt allain der Geometrj (versteh alles kunstlichen/gwisen/vnd vortailigen gebrauchs des Zirckels/Linials oder Richtscheittes vnd andrer werckzeüge/so zů allerlaj abmessen dienstlich) sonder auch der fürnemsten stuck vnd vortail der Rechenkhunst/furgeschriben vnd dargethon ist.

Auß Griechischer sprach in die Teütsch gebracht/aigentlich erklärt/Auch mit verstentlichen Exempeln/gründlichen Figurn/vnd allerlaj den nutz füraugen stellenden Anhängen geziert/Dermassen vormals in Teütscher sprach nie gesehen worden.

Alles zů lieb vnd gebrauch den Kunstliebenden Teütschen/so sych der Geometrj vnd Rechenkunst anmassen/mit vilfältiger mühe vnd arbait zum trewlichsten erarnet/vnd in Truckh gegeben/Durch

Wilhelm Holtzman/genant Xylander/ von Augspurg.

Getruckht zů Basel.

169

LE QVATRIESME LIVRE DES ELEMENTS D'EVCLIDE,

TRADVICT EN FRANÇOIS PAR Pierre Forcadel de Beziers.

DEFFINITIONS.

I

La figure rectiligne ſe dit eſtre deſcrite à la figure rectiligne, quant vn chacun des angles d'icelle figure qui eſt deſcrite, touchēt vn chacun coſté d'icelle figure, en laquelle elle eſt deſcrite.

FORCADEL.

Certainement ſi lon prent vn poinct en vn chacun coſté d'vne figure rectiligne, & on meine vne ligne droicte du premier au ſecond, vne ligne droicte du ſecód au troiſieſme, &c. Et vne ligne droicte du dernier au premier, lon aura deſcrit en icelle vne figure rectiligne, ſelon ceſte deffinition.

2

Semblablement auſsi la figure ſe dit eſtre deſcrite à l'entour de la figure, quāt vn chacun des coſtez d'icelle qui eſt deſcrite à l'entour, touchent vn chacun angle de celle, à l'entour de laquelle elle eſt deſcrite.

FORCADEL.

Certainement ſi lon fait paſſer des lignes droictes par tous les angles d'vne

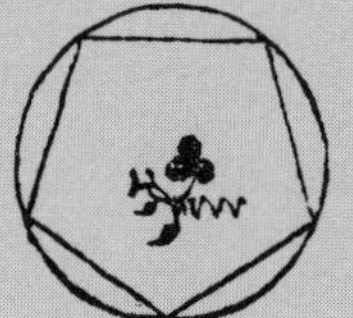

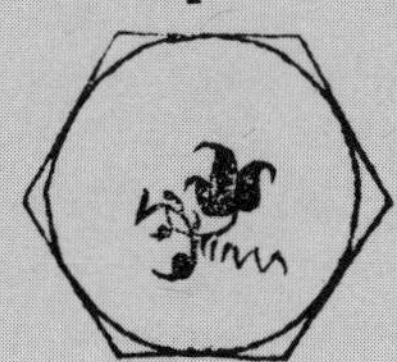

EE

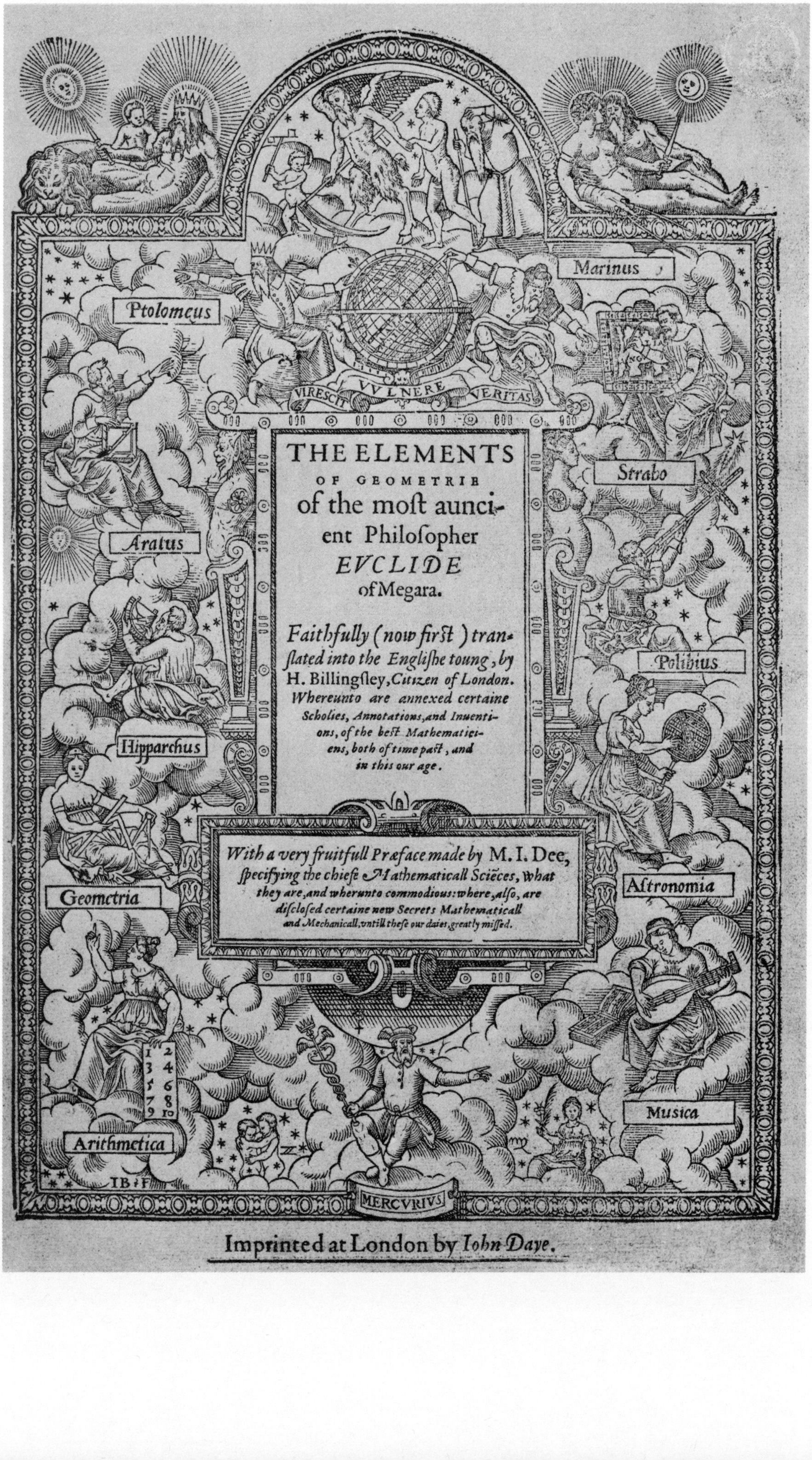

THE ELEMENTS
OF GEOMETRIE
of the most aunci-
ent Philosopher
EVCLIDE
of Megara.

Faithfully (now first) translated into the Englishe toung, by H. Billingsley, *Citizen of London. Whereunto are annexed certaine Scholies, Annotations, and Inventions, of the best Mathematiciens, both of time past, and in this our age.*

With a very fruitfull Præface made by M. I. Dee, *specifying the chiefe Mathematicall Scieces, what they are, and wherunto commodious: where, also, are disclosed certaine new Secrets Mathematicall and Mechanicall, vntill these our daies, greatly missed.*

Imprinted at London by *Iohn Daye.*

المقالة الرابعة وفيها ستة عشر شكلا

الحدود

اذا كان محيط دايرة يماس جميع اضلاع شكل مضلع او جميع زواياه او جميع اضلاع شكل مضلع يماس جميع زوايا مضلع اخر يقال للمحيط منهما انه مرسوم علي المحاط وللمحاط انه مرسوم في المحيـــــط ۞

الاشكال

آ

كل دائرة مفروضة معلومة لنا ان نرسم فيهـا وترا يساوي خطا مستقيما معلوما مفروضا ليس باطول من قطــــــرها ۞

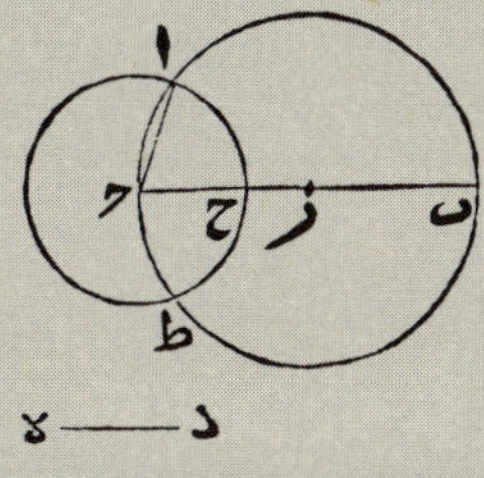

لبكن الدايرة آب ج والخط المفروض ده فنجد مركز الدايرة بالشكل الاول من الثالث وليكون نقطة ز ونرسم علي محيطهـا نقطة وليكن نقطة ب ونصل بينها وبين المركز بخط مستقيم ونخرجه في جهة ز الي ان ينتهي الي نقطة ج اعني محيط جانبها الاخر محيط ب ج قطرها فان كان الخط المفروض مساويا لخط ب ج فهــو المطلوب والا نفصل منه خطا يساوي خط ده بالشكل الثالث من الاولي وليكن هو خط ج ح ونرسم علي نقطة ج وببعد ج ح دايرة اح ط فيقطع محيطها محيط دايرة آب ج علي نقطتي آ ط ونصل بين نقطتي آ ج بخط مستقيم فهو يقع داخل دايرة آب ج بالشكل الثاني من الثالثة فلان خط ج آ يساوي ج ح وكان ده يساوي ج ح فخط ج آ يساوي ده فالحكم ثابت وذلك ما اردنا ان نبـــــين ۞

ب

كل دائرة مفروضة معلومة لنا ان نرسم فيهـا مثلثا يساوي كل واحدة من زواياه لنظيرتها من

Εὐκλείδου στοιχείων βιβλία ἕξ.

EVCLIDIS ELEMENTORVM GEOMETRICORVM LIBRI SEX, CONVERSI IN LATINVM *sermonem à Ioach. Camerario.*

Edebat Lipsiæ Georg. Ioach. Rhet.

OBELISCVS.

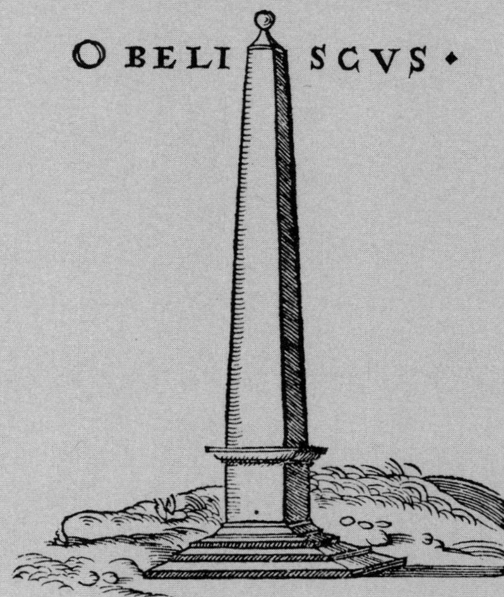

EXPRIMENTE VALENTINO PAPA. ANNO M. D. XLIX.

(*a*)

ΕΥΚΛΕΙΔΟΥ ΣΤΟΙΧΕΙΩΝ ΒΙΒΛΙΑ ιε΄.

EVCLIDIS ELEMENTORVM LIBRI XV.

ROMÆ. MDXXXXV.

Cum priuilegio S. D. N. *Pauli* III, *& inclytæ Reip. Venetiarum, ad quinquennium.*

(*b*)